THE POCKET CALCULATOR STORY

ANDREW MORTEN

AMBERLEY

First published 2024

Amberley Publishing
The Hill, Stroud,
Gloucestershire, GL5 4EP

www.amberley-books.com

ISBN: 978 1 3981 1686 3 (print)
ISBN: 978 1 3981 1687 0 (ebook)

British Library Cataloguing in Publication Data.
A catalogue record for this book is available from the British Library.

Typeset in 10pt on 13pt Celeste.
Origination by Amberley Publishing.
Printed in the UK.

Contents

Introduction

The story of the pocket electronic calculator is really the story of the technology that made it possible. All of the consumer electronic devices that we take for granted today, such as laptops, smartphones and flat screen TVs, exist because the technology behind them has been developing for many decades. In the 1960s, if you looked inside a television or a radio you would see warm, glowing valves, and later on, their smaller cousins – transistors. Transistors are at the heart of all electronic devices, and making them both small enough and cheap enough was the key to the electronics revolution that we benefit from today. Electronic calculators were the first mass-produced consumer product to make use of the advances in electronics in the late 1960s and early 1970s, and predate such devices as the video recorder and home computer by several years.

The first working transistor was produced in 1947 at Bell Labs in the USA. Three physicists – William Shockley, John Bardeen and Walter Brattain – shared the 1956 Nobel Prize in Physics for their achievement. The transistor can be used as a switch, and the electronic switch is the basis of the binary logic (on or off) that all calculators and computers use. Early calculating machines used mechanical switches instead where a handle had to be cranked to compute an answer. When transistors became commercially available in the mid-1950s, the car radio was one of the first devices they were used in when Chrysler offered a radio as an option in their cars produced from 1956.

A year later, in 1957, Sony launched the first mass-produced all-transistor radio, the TR-63, which went on to sell over 114,000 units. By the late 1950s, transistors had taken over from valves in most new electronics devices. Although the TR-63 was innovative, it only contained six transistors. To produce even a basic electronic calculator, however, required several hundred connected together in a particular pattern, along with hundreds of other electronic components. All of these had to be fitted to electronic circuit boards and wired together. The earliest electronic calculators were therefore large, expensive desktop machines.

The first steps towards miniaturisation happened in the US in the 1960s with the invention of the integrated circuit, or IC. The terms 'IC' and 'silicon chip', or just 'chip', are interchangeable and mean the same thing: a tiny slice of pure silicon containing various

An early metal-can transistor.

electronic components including transistors. Transistors are made from a material called a semiconductor, which has the properties of both a conductor and an insulator, and the manufacture of a chip uses a process called photolithography. In simple terms, the design for the chip is projected using ultraviolet light onto a piece of silicon that has been coated with light-sensitive material. The pattern is transferred onto the silicon when the non-exposed parts are removed with a solvent. Electronic components are built up in a series of layers by repeating the process many times. The result is a silicon chip containing, in the early days, hundreds of components, and nowadays millions.

In fact, the record for the most transistors on a chip is held by Apple for its ARM-based microprocessor, which contains 114 billion! This high density gives the necessary complexity that modern electronics requires, but also has advantages of lower costs and higher speeds of operation. Chips are made in semiconductor fabrication plants, or 'fabs', where they operate in super-clean conditions, many, many times cleaner than the cleanest operating theatre. Even a single particle of dust would ruin a chip if it were contaminated during manufacture. To be used in a device, the tiny piece of silicon then has microscopic-sized gold wires attached so that it can connect to the outside world, and then it is placed inside a ceramic or plastic package for protection.

Integrated circuits immediately found applications in military hardware, mainframe computers and the space race, where small, lightweight computers were being developed for the Apollo program. One of the first commercial applications was the electronic calculator. As well as compact electronic components, pocket-sized calculators would also need to have small displays developed and consume only the tiny amounts of power provided by batteries. These technical challenges were being tackled at the end of the 1960s and companies were starting to sense a new, potentially huge market in consumer electronics.

A naked silicon chip before being packaged for use in a circuit board.

A 1970s integrated circuit containing seventy transistors.

1

Basic Calculators

To be a useful tool, an electronic calculator needed to offer, as a minimum, basic arithmetic (add, subtract, multiply and divide) and also work with decimals. Other functions such as percentages, square roots and memories would come later, but initially manufacturers offered just a four-function calculator.

From the 1960s onwards, Japan and the US were to be rivals in dominating the electronic calculator market. In the US, companies had been funded by government spending on military and space, and in Japan, research was subsidised by their Ministry of Trade and Industry. This

A basic calculator – the Sinclair Oxford – from 1975.

competition came to a head in 1967 when Japan produced desktop calculators with ICs that infringed US patents on the manufacturing process. Japan needed the US market to sell them into, but the US threatened to seize the calculators at the ports. The standoff was eventually resolved when a joint venture between Texas Instruments (TI) and Sony was set up in Japan for the transfer of US technology. The technology that was being developed was called Large Scale Integration (LSI). This technique squeezed more and more electronic components into the chips, which brought down costs considerably. The fact that transistor density was increasing rapidly in integrated circuits led Gordon Moore of Fairchild Semiconductors to observe that the transistor count doubled every two years. This extraordinary increase in density has roughly stayed the same from 1975 to the present day and has become known as Moore's Law.

Manufacturers took advantage of this law during the 1970s to shrink electronic calculators down from desktop-sized to palm-sized devices before the end of the decade.

Calculator displays would also undergo a transformation during this period, from power-hungry glass tubes to solar-powered liquid crystal displays (LCDs). Electronic calculator costs were, almost unbelievably, in the hundreds of pounds for a desktop model in 1970, and in those days £100 was a lot of money! Fortunately, just a couple of years later, a pocket version could be had for just a few tens of pounds – still a great deal at the time and a similar outlay to purchasing a laptop or smartphone today. As the electronic calculator industry took off, the development of smaller and cheaper components continued at a fast pace. Calculators became more affordable each year, and therefore sales increased. This revenue fed back into the industry, with it being used for research and development into even smaller, cheaper calculators. By the mid-1970s, the basic calculator had reached maturity and become affordable to all as prices fell to just a few pounds.

This chapter describes some of the landmark calculators that were produced on that technological journey.

Anita 1011

Date: 1969
Features: Early integrated circuits, tube display, mains powered
Before the pocket-sized calculator had been developed, electronic calculators would sit on a desktop and require power from the mains. One of the last desktop calculators to be produced was the Anita 1011, made by a subsidiary of the Bell Punch Company of London. Bell Punch was founded in 1878 and originally made ticket punch machines, and later, mechanical calculators. Bell Punch was also behind the first desktop-sized electronic calculators when it released two machines – the ANITA Mk 7 and Mk 8 in 1961. The Anita Mk 8 resembled an old type of cash register and had over a hundred buttons on the front (ten buttons for units, ten buttons for tens, etc). It was also large and heavy at 13.9 kg (30.5 lb). But in 1966, the calculating side of Bell Punch Company was transferred to a new company called Sumlock Anita Electronics Ltd. In July 1969 they launched the Anita 1000 series of smaller desktop machines with ten-key keypads, and early models used a mixture of transistors and chips for the calculator logic. When launched in 1971, the Anita 1011 LSI (shown here), which was all integrated circuits, would have cost £250. It still required mains power and measured 9 inches wide × 4.5 inches deep and weighed 2 lb.

This Anita 1011 LSI dates from 1971.

It would have cost £4,000 in today's money.

As well as the basic four functions (add, subtract, multiply and divide) it has a memory and a per cent button. It uses a method of entering numbers called Reverse Polish Notation, or RPN, which means that it does not have an equals button. The first number is entered, then the [Enter 1st No] button is pressed, followed by the second number and finally the operator, e.g. [+]. The result is then displayed straight away.

The display is made from a series of tubes called 'Cold Cathode Numerical display'. They were often called 'Nixie' tubes, though this was a trade name of the Burroughs Corporation, which was an early developer of this technology. Nixie tubes were invented in 1955 and resemble the glass valves used in radios and televisions of the time, but in fact work in a different way. Inside is a wire mesh 'anode' and ten wire 'cathodes' in the shape of the numbers 0 to 9. The glass tube contains low-pressure gas and the numerals are made to glow when a high voltage is applied. Inside the 1011 there are ten Nixie tubes, allowing a ten-digit number to be displayed. Decimal points are displayed separately using small neon lamps.

Most of the calculating power of the machine resides in four chips, each of which contain about 200 transistors and were made by the Marconi-Elliot Microelectronics company. Marconi-Elliot was closed down in July 1971 by GEC during an industry downturn. Sadly, the whole of the British semiconductor industry eventually met a similar fate due to competition from the US and Japan.

Each tube contains ten wire-shaped digits.

Integrated Circuits (undersides)

Inside the Anita, showing the underside of the circuit board.

Sanyo ICC-82D

Date: 1970

Distinction: One of the first portable calculators

'Made in Japan' was a phrase that was synonymous with electronic gadgets in the 1970s, and one of the earliest Japanese companies was Sanyo. Dating from 1950, the Sanyo Electric Co. Ltd initially produced a plastic-bodied radio in 1952 and went on to produce many household appliances. Sanyo were an early pioneer of a chip manufacturing process called MOS (Metal Oxide Semiconductor) and were the first to use a type of logic gate called a shift register in a calculator. They launched their first calculator in 1966 and by 1968 Sanyo was using LSI (Large Scale Integration) chips from Motorola in the US to produce smaller calculators. At this time, in the late 1960s, Japan did not have the capacity to produce its own chips and so relied entirely on buying them from the USA. Companies in the US were ahead in designing and manufacturing LSI circuits, but Japan led in designing the calculators. Sanyo agreed a deal with the US General Instrument Corp (GI) to manufacture their design under licence, where GI received royalties, and fitted the Sanyo chips into their calculators. Later Sanyo had their own manufacturing facilities.

In September of 1970, Sanyo launched the ICC-82D, best described as a 'large, portable calculator'. Even though it was about the size of a tape recorder, it was actually the smallest calculator available (with a display) in the US at the time. The important

11

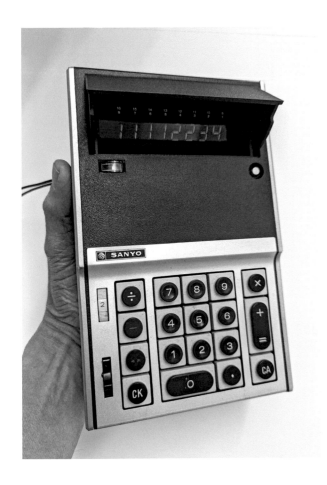

This early Sanyo calculator just about fits in the hand.

innovation was that it was battery powered and therefore truly portable. It measures 5.25 inches × 8.25 inches × 2 inches and, with its sturdy metal chassis, weighs in at 2.5 lb.

Inside the machine there are a set of Ni-Cad rechargeable batteries, which gave around three hours of use. At this time, around 1970, all of the transistors required for a basic calculator such as this, fitted into four chips. Many of the other components on the circuit board were needed to work the eight tubes which made up the display. These tubes worked differently to Nixie tubes, in that each digit was made up of seven electrodes. The electrodes could be lit independently to form whichever number was required. Each segment operated like a small neon lamp, so with about 60 volts applied, an amber glow was generated around that electrode. They were cheaper than Nixie tubes but required high voltage and high power to work them.

The 82D has a quirky feature not found on other calculators where the number of decimal points shown is pre-selected by a rotating switch. If the result of a calculation has more digits than the eight available, the carry lamp lights and pressing [< >] displays the surplus digits. There is also an error lamp if the calculation cannot be done; for example, divide by 0. It features a flip-up display cover (which turns off the calculator when closed), a battery-level meter, and came with a protective carrying case. These were expensive instruments at the time and cost the equivalent of £700 in today's money.

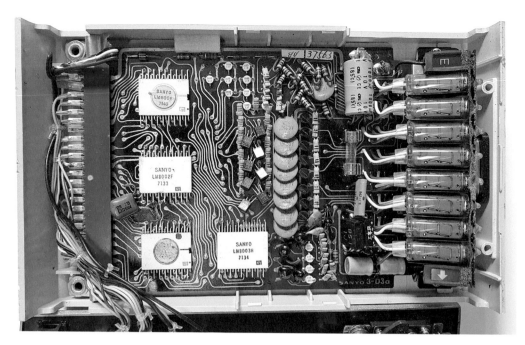

Inside, the four white chips provide the calculator logic.

The 82D was described as a 'mini' calculator at the time!

Sharp EL-8

Date: 1970
Features: Rechargeable battery and green, fluorescent display

The Japanese company Sharp was a major influence in design and development of early electronic calculators. The company dates back to the 1920s when it manufactured radios, and later on televisions in the 1950s. The name 'Sharp' is derived from the Ever-Sharp Pencil, invented in 1915 by Tokuji Hayakawa. The Hayakawa Electric Industry Company produced the first Japanese transistorised desktop calculator in 1964, the Sharp CS-10.

Sharp realised that one of the keys to reducing the size and cost of handheld calculators was in the area of the display. Nixie-type tubes were the main option around this time and not only were they bulky and power-hungry, but companies had to pay high royalties to Burroughs Corporation when they produced copies. Sharp got together with Ise Electronics Co. and invented the vacuum fluorescent display (VFD). These 'Digitron' tubes, as they were called, were used in a Sharp calculator by the end of 1967. The digits were made up of separate segments, as they still are today in such displays as digital clocks. However, the early VFD tubes used a different design of the digit to what became later the standard seven-segment display.

The 6s look strange on this early display.

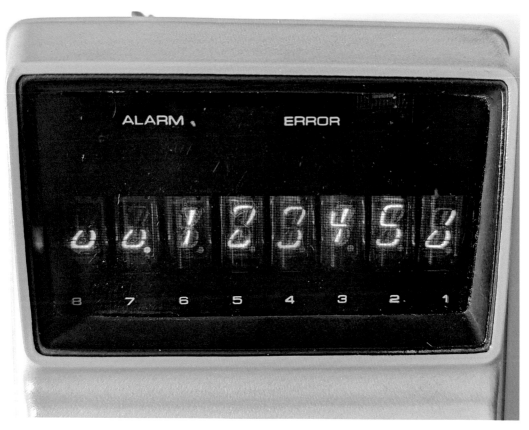

Displaying '00.123456' – an early design for digits.

Instead, eight-segment digits were used, but these made the 6 look odd, and the 0s look very small. Notice the segments that make up the sides of each digit are curved inwards. This meant that a 'conventional' 0, with its narrow waist, would have looked like an 8. For this reason, the 0 design just uses the lower half of the digit to avoid confusion. Also, the designers were unable to suppress leading 0s due to limitations of the electronics, so, for example, '1234' was displayed as '00001234'. Even though seven segments are enough to display all the digits from 0 through 9, early VFDs such as in the Sharp EL-8 used an extra eighth mini-segment to give a better looking 4.

VFDs work by attracting electrons to those segments that need to be lit to display the digit. These segments (anodes) are coated in fluorescent material which glows when struck by the electrons. The type of material defines the colour, typically blue or green, but others such as white and yellow came later. The electrons are given off a heated cathode, in the form of fine wires, in front of the segments and a suitable voltage between the anode and cathode makes the electrons flow – in many ways similar to a tiny cathode ray tube in an old television. The big advantage of a VFD tube over a Nixie tube is its size. A VFD tube contains one digit whereas Nixie tubes contain electrodes for all ten digits in a stack from front to back.

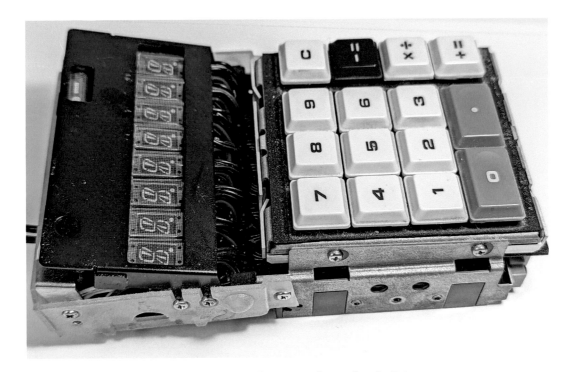

Without the case, the fine cathode wires can be seen in front of each digit.

The calculating board has been removed, showing the LSI chips.

With its new VFD technology, the Sharp EL-8 was smaller than predecessors at 4 inches × 6.5 inches × 2.7 inches, but still relatively heavy at 26 oz due to the rechargeable battery pack inside. It was definitely still a handheld and not a pocket calculator due to its thickness.

In order to minimise the number of buttons on the EL-8, multiply and divide shared a button. The [+=] button also multiplied two numbers together, whereas the [-=] button also divided two numbers. For example, 6 [X÷] 2 [+=] gives the answer 12 and 6 [X÷] 2 [-=] gives the answer 3.

The EL-8 was sold in the USA in 1970–71 and was one of the very first battery-powered handheld electronic calculators. In June 1971 it would have cost $345.

Sharp made calculators to be sold by other companies, such as Dixons in the UK, where they used the brand name 'Prinztronic'. For its work in the development of electronic calculators, Sharp Corporation was awarded the prestigious IEEE Milestone in Electrical Engineering and Computing in December 2005.

Busicom Le-120A 'HANDY-LE'

Date: Feb 1971

Distinction: First pocket calculator

The Busicom LE-120A, known as the 'HANDY-LE', is regarded as the first truly pocket-sized calculator, measuring 2.8 inches × 4.9 inches × 0.9 inches. Power was provided by four AA-size batteries. There were several models produced in the range, but it initially had a die-cast aluminium body which gave it a weighty and expensive feel.

The Busicom (Business Computer) Corporation of Japan changed its name from Nippon Calculating Machine Corporation (NCM) in the late 1960s. They were a very innovative company and were able to produce such a small device because, inside, all of the electronics required for calculations were fitted inside a single chip.

Busicom already sold a small desktop calculator called the NCR 18-15 or 'Busicom Junior'. They tasked circuit-design company Mostek of Dallas, Texas, to shrink the twenty-two LSI chips on two circuit boards of the Junior down to just a single chip. Initially it was not thought possible, but Mostek were desperate for the work and, after an estimate of six weeks turned into three months, they succeeded. In November 1970, Mostek announced they had produced the first ever 'calculator-on-a-chip', which they named the Mostek 6010L. They had managed to cram over 2,100 transistors from the original chips into a piece of silicon just 0.18 inches square – quite an achievement. The chip was immediately put to use in a new version of the Busicom Junior.

By February 1971, they had fitted the same chip into a new calculator – the 'Busicom HANDY' – the first pocket calculator. At $395, it was pocket-sized but not pocket-priced! The calculator made full use of the chip's ability to display twelve digits, though the number of decimal places was fixed at either 0, 2 or 4 and selected by a switch on the front.

The Busicom 'HANDY-LE' claims another first in that it uses LED modules for the display – hence the 'LE' in the name. LED stands for 'light emitting diode', and they are related to the transistor in that they are made from a semiconductor material but have two electrical connections instead of three. When current flows through them they emit

17

The first 'pocket-sized' calculator.

Inside, all the functions are performed by a single chip.

light of a specific colour. In the early days of their development this was limited to red, but nowadays any colour can be produced. LEDs were also expensive at the time, adding to the overall cost.

The HANDY-LE's high price meant that it didn't sell well, but Busicom had demonstrated that the technology had progressed to make a pocket calculator possible. The next challenge for the industry would be to drive down the cost and create a mass-market product. Busicom themselves ceased trading in 1974, but the name was bought by a British-based company and so the Busicom name continues to be used as a calculator brand. Being the first pocket calculator, the HANDY-LE is highly collectable and you would be lucky to see one for sale anywhere today.

Commodore C110 (By Bowmar)

Date: September 1971
Distinction: First US-manufactured pocket-sized calculator
Another milestone for the pocket calculator came in 1971 when the US started manufacturing their own calculators. The Bowmar company of Indiana manufactured LED displays in the late 1960s but when they could no longer sell them to Japanese calculator companies they decided to manufacture portable calculators themselves and get other companies to market them. Their first model released in 1971 was the 901B, for the very reasonable price of £79 ($200). It was also sold by Commodore Business Machines (CBM) as the C110, and by Craig as the 4501.

A Commodore C110, manufactured by the Bowmar company.

Inside there was another version of a calculator-on-a-chip. The announcement by Mostek in February 1971 that they had created their 6010 chip spurred into action fellow Dallas-based company Texas Instruments (TI). TI said that they were developing their own all-in-one chip to be available by June. The chip was not actually available until September 1971, but it was worth waiting for. Unlike the Mostek chip, the TI design was actually a microprocessor on a chip and included an arithmetic logic unit, a 182-bit random access store and a 3520-bit read-only memory for holding the program. The chip featured floating or fixed-point arithmetic, automatic round-off of numbers and leading-0 suppression. By changing the photo mask used to 'expose' the silicon at manufacture time, the contents of the program in ROM could be changed and therefore the functioning of the chip. The initial chip was named TMS1802, later renamed to TMS0102, and it was the start of a family of TMS01xx chips. By November 1972, TI offered nine different calculator-on-a chip circuits to calculator manufacturers at around $40 each for a quantity of 100 pieces. The CBM C110 used the TMS0103 version, which gave the calculator four functions with an eight-digit display. It can be seen from the date code A7211 in the photo that this chip was manufactured in November 1972.

Measuring 3 inches × 5.2 inches × 1.5 inches, it is more handheld than true pocket-sized due to the six rechargeable AA-size batteries inside. The keyboard used is called 'Klixon'

Inside is the Texas Instruments TMS0103 calculator-on-a-chip.

by Texas Instruments and gives a very positive feel; it was used in many calculators of this period. Here it is missing a separate [=] button, as was common on early calculators. The [+=] button is used instead unless one of the operands is negative in which case [-=] is used.

When the 'K' is set to 'on', a previously entered number as part of a multiply or divide calculation is remembered, and subsequently just pressing [+=] uses this number in the next calculation.

The LED display has eight digits, plus a ninth that indicates either 'E' for an error, 'L' for low battery, a square for overflow or a dash for a time out. The display turns off after fifteen seconds of non-use to save power, and the D button turns the display back on. At £79 it was the cheapest calculator you could buy in Britain at the time.

Bowmar were very successful and made a range of calculators up until the mid-1970s. During the calculator 'boom', when prices were falling fast, companies were having to bring out new features all the time and components were scarce. Bowmar failed to keep up and they filed for bankruptcy protection in 1976. They struggled on, however, shifting into specialised mechanical design, including work on the Apollo space program, and were acquired by White Electronic Designs. In 2012, the Bowmar name was brought back when their original plant in Fort Wayne, Indiana, was bought by an investor.

A later Bowmar model – the 90911 from 1974.

Rapid Data Rapidman 800

Date: February 1972
Features: Single chip design, price under $100

By 1971, the race was on between companies to drive down the price of electronic calculators. They wanted to break the $100 price barrier and create a high-volume consumer product. Components only accounted for around a quarter of the selling price and therefore the $100 calculator would have to source everything (chips, display, case and keyboard) for no more than $25. In early 1971, the four-or-so MOS chips required for a basic calculator alone cost around $40. A few months later, several companies – such as Texas Instruments, Mostek, General Instrument and Caltex – were suppliers of one-chip calculator circuits priced at around $12–15, but other LSI manufacturing companies announced the intention to build them for only around $5–6 each. The sub-$100 calculator was now in reach if companies could keep the rest of the component costs down to an absolute minimum.

One of the leaders was Rapid Data Systems & Equipment Ltd of Canada. The company produced several innovative calculators in the 1970s, the first of which was the Rapidman 800. It measured just 3.15 inches × 5.4 inches × 0.9 inches and, at around $99 at the launch, it can claim to be the first affordable pocket calculator. 'Pocket' is a relative term of course, but like smartphones today that are getting bigger and bigger, the Rapidman could just about get away with the claim despite being nearly an inch thick. The low cost was

The Rapidman 800 by Rapid Data Systems.

achieved by a couple of innovations. Firstly, they produced an injection-moulded plastic case which incorporated the front, back, all the keys and battery cover in one piece. This cut assembly costs as there was no keyboard to put together, but the drawback was that the calculator had a cheap, mass-produced feel to it.

The second cost-saving feature was the use of a single chip to perform all the calculating functions. After Mostek had designed the 6010L chip for Busicom in late 1970, they released a chip for general use, designated the 5010, and this is what Rapid Data used in their Rapidman 800. The silicon chip itself sits on the white ceramic package, which has gold 'legs' along two sides to make the electrical connections. This packaging style is known as a dual-in-line (DIL) from the arrangement of the legs.

Although all the logic required for the calculator was now housed in the Mostek chip, there were some other components – resistors, capacitors and transistors – required to drive the display. Light-emitting diodes (LEDs) were increasingly being used in 1972 as they were becoming cheaper through mass production. Here, three three-digit modules are used to give nine digits in total. As early LEDs were very small, lenses were fitted over the digits to magnify the numbers and improve readability. Despite their relatively high power consumption, the larger PP3 (9V) battery gave a reasonable battery life. The Rapidman 800 may not have been the first, but it was one of the very first affordable pocket calculators that sold in large numbers due to breaking the $100 barrier.

The Rapidman case was a one-piece to cut costs.

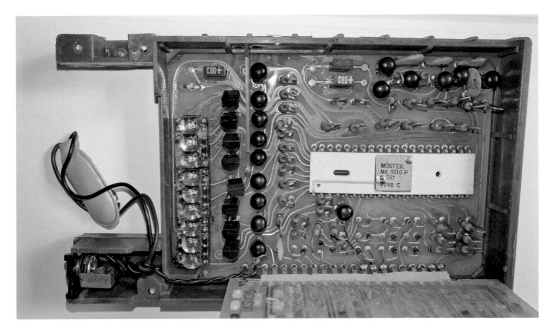

The gold-coloured chip performs all of the logic for the calculator.

Casio Mini

Date: August 1972
Features: Low cost, six-digit display

The stated aim of the Japanese electronics industry in the late 1960s was to capture as much of the world's mechanical calculator market as possible with their electronic replacements. By 1972, they were well on the way. The huge increase in demand for LSI devices from Japanese calculator manufacturers caused prices from the chip factories to plummet to below US prices and Japan now bought 50 per cent of chips from home-grown semiconductor manufacturers. For the moment, the US had lost any technical lead it had in chip design, at least as far as the more basic designs. Now they would have to compete for customers on the basis of price. One of the Japanese companies taking advantage of this boom was Casio.

The Casio brand is synonymous with digital watches, electronic keyboards and, of course, pocket calculators. They produced an electro-mechanical desk calculator as far back as 1957, and in the 1980s and 1990s Casio were at the forefront of innovations in quartz wrist watches, digital cameras and scientific calculators. In 1972, they produced one of the first affordable handheld calculators: the Casio Mini, for $59.95.

To help keep the price low, the display was limited to six digits rather than the normal eight or ten. Each digit was produced by a green vacuum fluorescent tube rather than the more expensive LEDs. Inside, a Hitachi-made chip provided the logic for the four functions and four AA-size batteries give the Mini a useful battery life of around ten hours. An unusual feature is that the Mini has no decimal point button. If the result of dividing one number by another is a fraction, only the whole number is displayed. To see the fractional

The Casio Mini displaying '123450'.

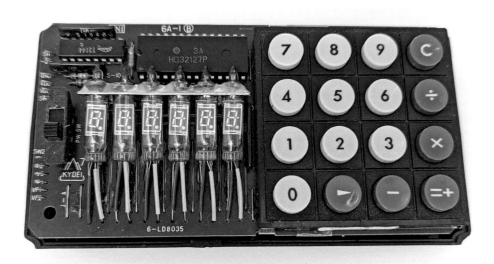

Six tiny VFD tubes display the numbers.

part after the decimal point, the right arrow key is used to scroll up to a further six digits onto the display. Despite this apparent limitation, the Mini was very successful and sold in the tens of thousands. The Mini was popular in Japan for currency calculations where the lack of a decimal point didn't matter. The design was tweaked in a couple of later models to lower the price further.

Casio's track record in producing quality, innovative products and being able to read and anticipate the fast-moving world of consumer electronics ensured their future success.

Texas Instruments TI-2500 'Datamath'

Date: July 1972
Distinction: Texas Instruments' first calculator
Texas Instruments produced a sophisticated design for a calculator-on-a-chip in 1971 and went on to release a whole family of related chips, named TMS01xx – the 'xx' denoted the chip's model number. Each chip in the family had slightly different functionality according to requirements of the calculator it would be fitted into, such as a memory option, the number of digits and floating point options. Less than twelve months after developing the TMS series, the integrated circuit company released their own calculator using one of these chips, the TMS0119. It was well received by retailers in the US, who thought it was better looking than the Japanese offerings at the time, and the price of $149.99 was low enough to attract customers in large numbers.

Measuring 3 inches × 5.5 inches × 1.7 inches, it is chunky but comfortably handheld. Most of the internal volume is in fact taken up by the six AA-size rechargeable batteries. There is an eight-digit red LED display and four functions. An extra feature is the Chain/Constant switch above the [C] button. If set to 'Constant', the calculator remembers the second operand in a subsequent calculation. For example, if the sum '1.2 × 2.4 =' is entered, and the next sum is to be '10 × 2.4', just entering '10' and pressing [=] gives the answer '24'.

The Datamath model from Texas Instruments.

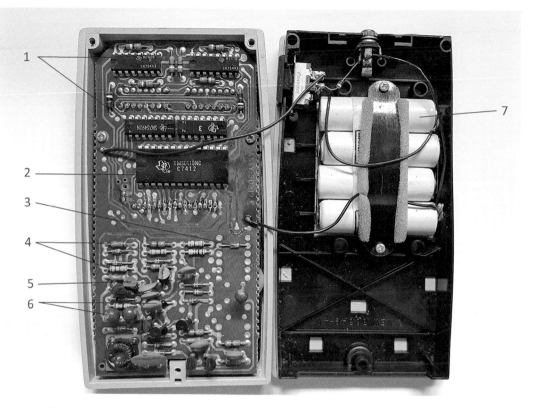

Inside a Datamath, highlighting the main components. **Key:** 1) Driver chips that control the LED display. 2) TMS0119 chip which provides all the calculator's functions. 3) Glass diode. These restrict the flow of current to one direction. 4) Resistors. Coloured bands denote their value in ohms and are used to regulate current. 5) An individual transistor in a plastic package used in the calculator's power circuitry. 6) Two different types of capacitors. 7) Rechargeable battery pack.

The TI-2500, known as 'Datamath', was a success and the first in a long line of handheld and pocket calculators by Texas Instruments.

Sinclair Executive

Date: August 1972
Distinction: 'Smallest Calculator In The World'
The Executive from company Sinclair Radionics was a radical departure from the standard boxy calculators of the time. The design was the most important aspect of the Executive from the beginning. The brief was for the calculator to be small, slim and tactile. Its proportions were not so different from a modern smartphone, but this was thirty-five years before the iPhone was designed.

Sinclair Radionics (a contraction of radio and electronics) was started by Sir Clive Sinclair in 1961. The company specialised in radio and hi-fi and particularly in designing

A matt-black Sinclair Executive.

products that were miniaturised. The first of these was the Sinclair Slimline radio in 1963, followed by the Micro-6 matchbox-sized radio in 1964. The Micro-6's claim to fame was as the 'world's smallest radio'. Sinclair was constantly trying to push the boundaries of what was possible with the technology available at the time. Inevitably, sometimes he pushed too hard and some of his amplifiers and stereos were commercial failures. However, another of his radios, the 'Micromatic', was a success and sold from 1967 up to 1971 as an even smaller 'world's smallest radio'!

In 1972, Sinclair Radionics turned from audio to calculators and entered the market with a product that met their trademark claim of 'the smallest in the world'. Quite a claim – but the Executive was just that as measured against other calculators of the time at just 5.5 inches × 2.25 inches × 0.375 inches. The advertising said it was as thin as a cigarette packet, a sign of the times and perhaps a reference to Sir Clive's smoking habit. The intention was that, like a packet of cigarettes, it could be carried in a shirt pocket.

The calculator buttons are standard, except for a couple. First, the switch above the buttons is called [K] and remembers a number to be used as a constant in calculations. The [-] and [+] buttons share with [=], so display the result as well as subtracting or adding. There is also a slider switch on the right side to select either two, four or six decimal places or a floating decimal point (with no rounding).

Sinclair's 1967 Micromatic radio next to a 10p piece.

It was acclaimed at the time for being stylish – almost a piece of art. An article in the *New Scientist* journal written by Nicholas Valery telling of his experience in owning one said: 'It's not so much a professional calculator but more a piece of personal jewellery. And as such it continues to delight and amaze people.' This reduction in size was not achieved through any sudden technological breakthrough but by clever, innovative design.

First, the Executive used calculator-on-a-chip design, initially using the Texas Instruments TMS1802. This 7000-transistor microchip was described as the calculator's 'brain' and the largest ever produced for a commercial product. The buttons were miniaturised and each sits over a tiny spring used to press a copper contact onto the circuit board. Switches for on/off and decimal places were moved to the calculator's edges. Inside, a novel technique is used to reduce the power consumption of the hungry TI chip. Sinclair discovered that there was a latency in the display and electronics, i.e. they stayed 'on' for a time after power was removed. He used this fact to create a timing circuit that switched the power on for just 1.7 microseconds in each 65-microsecond period. Amazingly, it worked, even though Texas Instruments didn't approve of their chips being used this way and warned against potential problems.

Another power saving technique was to drive the display directly from the battery rather than using buffer chips. Together, these reduced the power consumption from 350 mW down to just 20 mW. This reduction allowed Sinclair to shrink the batteries and use the button cells normally found in hearing aids. This was the big step in reducing the size and

The Executive is just 8 mm thick.

weight of the calculator. Battery life wasn't great at only about four and a half hours, but it was thought to be a worthwhile trade off in producing a world first in miniaturisation.

Apart from the short battery life, reliability was also an issue, at least in the early days, but Sinclair was so confident in his product that he offered a 'no quibble' five-year guarantee to allay people's fears. Even so, it was reported that a Russian diplomat who owned one had it explode in his pocket, which allegedly lead to an official Soviet investigation. The cause was found to be due to the batteries overheating and bursting after being accidently left on for a long period.

These teething troubles led to evolution of the design over its lifetime. The large LED display was replaced by a Bowmar display with smaller numbers but with a lens in front of them to make them readable. This improved battery life but at the expense of reducing the viewing angle.

The keyboard underwent a minor change in what became the 'Type 2', which replaced the [K] switch with a High/Low switch for the display brightness setting. The [K] now had its own button. The chip was also changed to a GIM (General Instrument Microelectronics) C550 and the decimal point switched was removed. The battery life was now more than doubled to nine and a half hours.

An updated model, the Sinclair Executive Memory, was released in 1973. The number of buttons on the keypad was kept the same and the slider button, which had originally been used to remember a constant (K), was now labelled [Σ] meaning sum, and used to place a number into the memory. Recall from the memory was by the [M] button on the keypad bottom left. Internally the TI TMS0132NC was used along with the addition of a couple of

Executive's box proudly claims 'World's Smallest'!

Two versions of the Executive Memory.

driver chips for the display. Externally, the Executive Memory was available either in white, or black with a distinctive gold-coloured keypad.

When launched in 1972 the Executive sold for £79.95 plus VAT – comparable to purchasing a high-end laptop or iPhone at today's prices. The price was reduced to £59.95 in 1973 to boost sales, which worked, and by 1974 Sinclair were producing 100,000 a month, many for export. The Executive was a big financial success for the Sinclair company as well as being a design classic and made the company £1.8 million in profits.

The Executive won its designer Richard Torrens the Design Council Award for Electronics in 1973 and it was displayed at such venues as the Science Museum in London and the Museum of Modern Art in New York.

Sinclair Cambridge

Date: 1973
Features: Small size, available as a kit
The Sinclair Cambridge range of calculators were characterised as being small and affordable. Whereas the Executive used button cells, the Cambridge used conventional AAA batteries to achieve a reasonable battery life. They therefore weren't as slim as the Executive but had a smaller outline at just 2 inches × 4.4 inches. The small LED display was from pioneering US manufacturer Bowmar. This was a very compact calculator indeed at the time and weighed in at less than 2 ounces without the batteries.

The manufacturing costs were kept low by using cheaper components, which enabled Sinclair to initially sell them at under £33, and a kit was available for £5 less. The affordable prices boosted sales and made the Cambridge a success. On the downside, the low quality on/off switch had tin-plated contacts which wore out quickly and was the Cambridge's Achilles heel. There were nearly a dozen Cambridge models over their lifetime.

The first models, called Type 1, Type 2 and Type 3, were all basic four-function calculators. The Type 1 has a brown glossy case and a [K] (constant) button. The Type 2 has the same case but no [K] button, but instead has automatic constant on the second operand. The Type 3 has a black matt case, a reinstated [K] button and a [C/CE] button to clear the last entry or clear everything. A Cambridge Type 4 model was released in 1975 where a new microchip, the General Instruments G-595, gave it a per cent function. It also required less power and so only needed two batteries instead of the four in previous models.

From 1974 to 1976, a range of Memory models were produced. The first, the Type 1, has a black case and a single button labelled [RM CM] for memory recall and clear. The Type 2 version has a silver case and the addition of a percentage function. The final Type 3 version has an orange case and takes a 9V PP3 battery instead of the AAAs of previous models to extend the battery life. However, the PP3 size didn't fit inside the battery compartment and so the back of the case bulged outwards, causing it to become known as the pregnant model!

The final version, the Universal, was produced up until late 1977. This also has a memory and percentage button but has the addition of square, square root and reciprocal functions.

All models came with a protective case and instructions which had examples of how to perform currency conversions, mortgage interest repayment and compound interest.

A Cambridge Type 4 model.

Left to right: Cambridge Type 2, Memory Type 1 and a Memory Type 2.

A Cambridge Universal model and a brightly coloured Memory Type 3.

Toshiba BC-0808

Date: 1973

Feature: Single-tube vacuum fluorescent display

The Japanese company Toshiba is a multinational technology corporation with headquarters in Tokyo. The name comes from a contraction of TOkyo SHIBAura Electric Co., and its history can be traced back as far as 1890 to a company that made light bulbs. Today its business interests are vast, ranging from home appliances and medical equipment to personal computers and semiconductors. But from the late 1960s to the mid-1970s they manufactured desktop and handheld calculators. Their first handheld calculator was the BC-0801 in 1972. It measured 4 inches × 6.5 inches × 1.5 inches and required six AA-size rechargeable batteries. A year later they produced the BC-0808, which was similar in design but had a lower electronic component count and required only four AA-size batteries.

The display used in both calculators is a second generation vacuum fluorescent display (VFD). In the first generation VFD each digit was housed in a separate tube. Second generation VFDs packed all eight digits into a single tube, saving both space and cost. Before LCD displays were commercially available in the mid-1970s, calculators used both LEDs and VFD displays. Each had its pros and cons. Early LEDs were small, so made for a compact calculator, but the display was difficult to read for the same reason. They were also expensive. VFDs were larger due to the bulky glass tube and they required a high current

This BC-0808 was a promotional gift from the brewer Heineken.

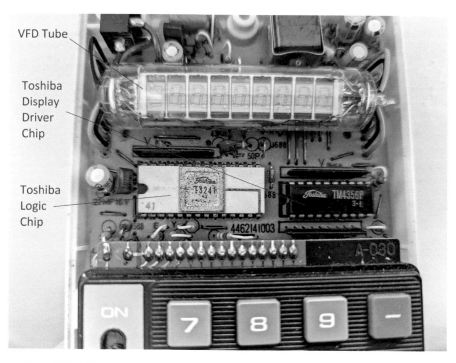

Inside a BC-0808.

source to drive them. Both types were used in pocket calculators until LCDs replaced them. Incidentally, VFDs are still widely used today in products such as hi-fi and digital clock displays, where the display needs to be visible in all light conditions but where power consumption is not an issue. The logic chip used in the BC-080X series was a TI design but manufactured by Toshiba, and performed all of the calculator logic functionality.

Royal Digital 5T

Date: 1973

By 1973, the calculator market had changed dramatically. In the two years since Japan had caught up with the US in production, prices for calculators had fallen from around £125 to around £25 in late 1973. At the same time, Japan's share of the world market fell from 80 per cent to around just 40 per cent. Japan was still investing heavily and producing 10 million units a year, but the dramatic fall in prices hit profits. Meanwhile, in the US the picture was much brighter. Companies that started out making components, such as Bowmar, had moved into calculator manufacture and were doing very well. Texas Instruments, the world's largest chip manufacturer, had also successfully diversified into calculators and had opened a new factory in the UK in Plymouth. Hewlett-Packard had taken the lead in specialised financial calculators, and Rockwell International, which had been prime contractor on the Apollo programme, took over Britain's largest calculator manufacturer – Sumlock Anita.

The Digital 5-T from British company Royal.

Inside, showing the blue fluorescent display tube.

There were still a few British calculator manufacturers hanging on in 1973, and one of these was the Imperial Typewriter Company Limited of Leicester. Dating back to 1908, they were the household name for mechanical typewriters in the 1960s. In 1966, they merged with Royal Typewriters, owned by Litton Industries, and designed electronic calculators to be assembled in the far east. In all, they marketed around twenty-five different models under the Royal and Imperial brands up until the mid-1970s.

The Royal Digital 5T is typical of the range in its boxy design, measuring 3.5 inches × 5.9 inches × 1.4 inches, with an eight-digit vacuum fluorescent display. It takes four AA-sized batteries with a claimed four-to-five hours of use, or rechargeable cells could be fitted as a charger was supplied with the calculator.

After a dispute over pay and conditions in 1974, the Leicester factory closed down. The sales office was renamed Imperial Business Equipment and continued until 1989.

Rockwell 8R

Date: 1975

The Rockwell conglomerate in the US could trace its history back to 1919, before it was split up in 2001. Its main businesses were aerospace, machinery, defence and electronics.

In the late 1960s, Rockwell's Autonetics division was a leading player in manufacturing Large Scale Integration (LSI) ICs and supplied calculator-makers such as Sharp. By mid-1972, Rockwell was making and supplying its own calculators with LCDs to companies such as Rapid Data and Sears Roebuck to be sold under their own names. Using Rockwell components, the cost to the consumer was kept down to the $100 mark. Success led

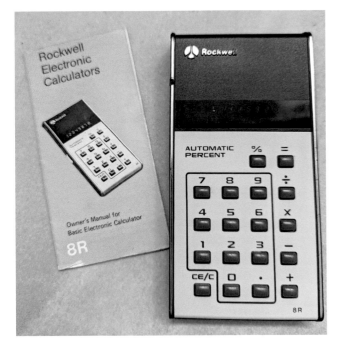

The 8R from US company Rockwell.

Inside the 8R are very few components, keeping the costs low.

Rockwell to buy US company Unicom in 1972, a calculator manufacturer and distributor, and then acquired the British calculator manufacturer Sumlock-Anita in 1973.

But the calculator boom for Rockwell was short-lived. After sales peaked at around $130 million in 1976, competition from Japan forced Rockwell to quit the calculator business completely and concentrate on aviation and space.

The 8R model is typical of the Rockwell range of calculators sold up until the mid-1970s. It requires a 9V battery and offers a per cent function along with the standard four functions. The number of components inside calculators continued to fall and the 8R has just five – the display, on/off switch, two capacitors and the logic chip. They were manufactured widely at locations including Mexico and the UK. They sold in huge numbers and can still be obtained cheaply today on auction sites.

Casio Pocket-LC

Date: 1975
Distinction: First calculator with a modern LCD display
In the early 1970s, the type of display used by all calculators was either LED or vacuum fluorescent display (VFD). Both of these were fairly power-hungry and therefore shortened

the calculator's battery life considerably. In 1972 a form of liquid crystal display (LCD) called DSM (dynamic scattering mode) was invented by Rockwell and further developed by the Sharp Corporation. DSM LCDs go from transparent to opaque white when a voltage is applied. However, the display required to be backlit for it to be visible and so needed a filament bulb behind. Early calculators that used these LCDs were not very successful due to the power required for illumination. There were also reliability issues.

In 1975, a second-generation LCD was developed which solved these problems. The Twisted Nematic (TN) LCD, as they are called, displays black numerals on a grey background and so does not require any background lighting (though obviously are not visible in the dark like LEDs and VFDs are). More importantly, the new LCD consumed negligible power and so was ideal for a battery-powered calculator. Early LCDs were sensitive to ultraviolet light, which could degrade them, so were fitted with a yellow filter. Casio was the first company to sell a TN LCD calculator – the Pocket-LC in 1975. Being powered by two button cells, the calculator could be shrunk in size to just 2.5 inches × 4.3 inches × 0.5 inches. In addition to the standard four functions, there is a memory and per cent function.

The success of the Pocket-LC was followed a year later by the Pocket-LC II. The only significant change was that the button cells were replaced by a single AA-size battery. This meant that the calculator's size increased slightly but gave a claimed battery life of 250 hours, which was unheard of amongst pocket calculators of the time.

Casio's first LCD calculator.

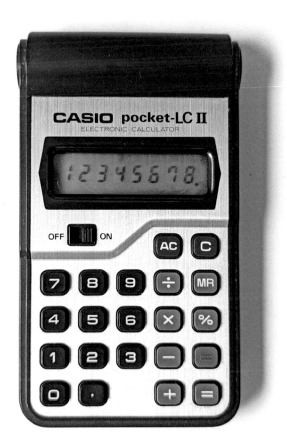

The later Pocket LC-II model, which had a battery compartment at the top.

Commodore 776M

Date: 1976
Feature: Low cost

By November 1975 the market was being flooded with calculators, from the simple four-function models for about £5 to scientific models for around £50. In the first five years of the 1970s, handheld electronic calculator manufacturing went from 0 to 50 million units and created an industry worth $2500 million.

But 1975 was the year of reckoning for the industry. Prices crashed, and with them went company profits. Any company not big enough to weather the storm would fail. It had been fashionable just a couple of years earlier for companies to make all of the components themselves, even the chips, by either opening new factories or buying out other companies. It was known as 'vertical integration', where the calculator manufacturer controlled everything and ensured there was always a plentiful supply of components. But having so much tied up in calculator production, even chip manufacture was tough when profits in the industry were suddenly hard to come by.

One of these companies that was struggling was Commodore (known as CBM in the UK), which made a $4.3 million loss in 1973 even though its sales were up by 12 per cent. But in a move that gave Commodore complete integration in the calculator business, it acquired

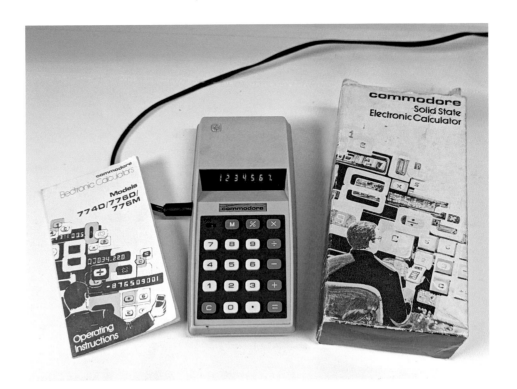

A popular 776M model from Commodore.

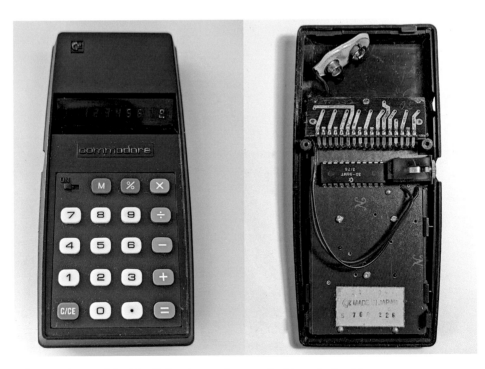

A later 796M model with all the electronics now inside a single chip.

the chip manufacturer MOS Technology Inc. of Pennsylvania to integrate downwards in its supply chain. Commodore had already worked with MOS in supplying chips for other calculators in its range, such as the SR7919 model, which was popular in the UK. Another popular low-cost model by Commodore was the 776M. It was a basic calculator but with the addition of per cent and memory functions. Inside the case, apart from the seven-digit display, there are just two chips. One for the calculator logic and one to drive the display.

The following year CBM launched a similar model – the 796M. The only real difference was an eight-digit display rather than seven. Inside, there was now just a single chip – a 3D-98MT which could drive the display directly. Both these models were popular and sold well due to their low cost.

Buying MOS Technology turned out to be a very shrewd move for Commodore because, as the calculator market matured in the late 1970s, Commodore used its chip manufacture business to make microprocessors and other computer chips as it moved into the then fledgling, but soon to be booming, personal computer market.

Sinclair Sovereign

Cost: £30
Feature: Stylish design

The Sinclair Sovereign model introduced in late 1976 was named for Elizabeth II's Silver Jubilee the following year. By this time there were many basic calculators available cheaply and multi-function scientific models were also reasonably priced. So, to introduce yet another power-hungry LED model was a questionable business decision. Sir Clive decided to go upmarket and produce a relatively expensive, stylish model he thought would appeal to more affluent buyers.

John Pemberton was the designer; he came up with a very sleek design just 1.4 inches wide, 5.6 inches high and 0.5 inches deep. The case was made of pressed steel, which uniquely allowed the calculator to be produced with different metal finishes. As well as matt black, there was brushed chrome, bright chrome and gold plate. Prices ranged from £30 to £60, and a limited-edition silver-plated version was made in 1977 with Elizabeth II's Silver Jubilee crest on the front. A few were reportedly made in solid gold for the richest of customers.

Some were ordered by companies as gifts for clients and had the company's logo embossed on the case. The quality feel was extended to the packaging which was either a solid plastic case with storage for the soft sleeve, batteries and manual, or a wooden box for the gold-plated version. John Pemberton won the Design Council Award in 1977 and there are examples in the Museum of Modern Art in New York.

Functionally, the calculator has useful functions beyond the four for arithmetic. There are five memory functions (add, subtract, exchange, clear and recall), square, square root, reciprocal and per cent. This is all achieved with the single Mostek MK50321N chip.

The Sovereign had limited success beyond the novelty of its classy exterior. Its Achilles heel was, as per the Executive model, its battery life. Sinclair's claim of the batteries lasting weeks or months relied on customers switching them on for a few seconds a day to make a calculation before switching them off again. This wasn't how customers actually used them, and so the two batteries lasted only hours before needing to be replaced.

This matt-black Sovereign was a gift from the Plessey electronics company.

A silver-plated version for Elizabeth II's silver jubilee in 1977.

A gold-plated version.

Two 1970s adverts from the *Sunday Telegraph*.

A number of factors, such as high import duty on electronic components, reliability issues and the fact that other manufacturers were moving to low-power LCD displays, meant that 1977 marked the end of the line for Sinclair and calculators. However, Sinclair did carry on with electronic innovations when they developed a range of successful home computers in the 1980s.

If you can find a Sovereign today they sell for high prices in good condition, though the plating was thin and so the silver or gold is often worn from over-polishing.

Casio Micro-mini

Date: 1976

Features: Matchbox-sized with LCD display

Calculators with a single chip inside to perform all of the logic had been around since 1971 when the Busicom LE-120A was launched. Although the actual 'chip' of silicon didn't get much smaller over time, the packaging did. Originally the chip was attached to a large rectangle of ceramic material with the electrical connections along two of the sides in a dual-in-line (DIL) package. By 1976, other more compact packaging had been designed, such as that used in the Micro-mini. Here, the chip is housed in a square plastic package with legs on all four sides, known as a 'Quad Flat Package', which takes up much less room than DIL. Another advantage of this type of packaging is that the legs can be soldered directly onto the surface of the circuit board instead of passing through the board and being soldered underneath. This means the circuit board is simpler and costs less to assemble.

The diminutive Casio Micro-mini.

Space for button cell battery

Power supply circuit

Integrated Circuit in 'Quad Flat Package'

Inside a Micro-mini.

With the new chip packaging, coupled with a low-power LCD display powered from a single button cell, it was technically possible to produce a tiny calculator less than 2.5 inches long. This is what Casio did with their Micro-mini models, and claimed they were the smallest in the world at the time. This model is the M810, which has an extra per cent function and a memory over the earlier M800 model. LCDs still appeared yellow around this time as they were affected by UV light and required the yellow filter.

Although it was possible to produce a matchbox-sized calculator, it wasn't necessarily a good thing. A conventional-sized pocket calculator would still fit in your pocket or handbag, but its larger size meant that the buttons were useable. On the Micro-mini, the buttons are too small to be practical. It would also not fit comfortably in a wallet due to its thickness, so the life of the micro-mini-sized calculator was limited.

Texas Instruments Programmer

Date: August 1977
Feature: Calculator for computer programmers
At first sight, Texas Instruments 'Programmer' model looks more like a scientific or programmable calculator, but it actually just has the four arithmetic functions. Its

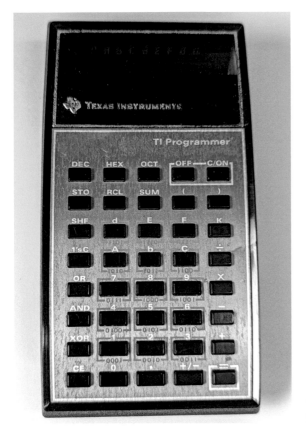

The Programmer displaying hexadecimal number ABCDEF00.

speciality is that it also works in base 8 (octal) and base 16 (hexadecimal) as well as decimal. The TI programmer was designed specifically for computer programmers, as a computer's native number system is base 16, and it is common to need to covert base 10 to 16 and vice versa. TI famously advertised it as 'a calculator that's of very little use to 99.9994 per cent of the population'!

Hexadecimal has the extra letters A, B, C, D, E and F to represent numbers 10 to 15. For example, to convert hexadecimal F to decimal, pressing the [HEX] button puts the calculator in hex mode, and pressing the [F] button then the [DEC] button displays the answer 15. An inverted comma or quotes mark on the display indicates which base you are in. Hexadecimal numbers are represented in binary by four bits or binary digits. Although the blue writing underneath the number buttons is their binary equivalent, the calculator does not offer binary conversion – that would come later. It also has a memory and some buttons on the left-hand side called logical operators, used for manipulating individual bits of a number. There is OR, XOR and AND which work with a pair of numbers, shift [SHF] which moves all the bits of a number along by one, and 1's compliment [1'sC] which inverts all the bits in a given number.

The calculator has an eight-digit LED display, is powered by a 9V battery and cost £49.99 at launch. After 30 seconds of non-use, the display shows an animated dot running across the display to show it is idling.

Sharp EL-825

Date: 1980

Feature: Early solar-powered calculator

As the pocket calculator market matured further at the end of the 1970s and prices hit rock bottom, manufacturers had to continually improve designs in order to sell new models to the public. Not everyone wanted a scientific or programmable model, but everyone wanted their calculator to be reliable, small and with a long battery life. It was better still to remove the battery altogether. The earliest calculator featuring solar cells was Sharp's EL-8026 in 1976, but the cells were on the rear and could only charge the batteries when the calculator was switched off. The EL-825 released in 1980 though is truly solar powered – there are no batteries at all. It features an eight-digit LCD display, four functions plus per cent, memory and square root. The calculator has a unique hinged design, so when it is opened up the solar cell, which is the same size in area as the calculator, lies flat and provides enough current to power the calculator under normal room lighting. When folded up it is a little bigger than a credit card outline. Solar cell technology had improved significantly around this time, and it was possible to shrink the solar cell down to a few square centimetres, rather than use the rather big cell on the EL-825. However, the larger cell meant that it would work in low-light conditions.

A Sharp EL-825 with its large solar cell.

2

Scientific Calculators

As the basic electronic calculator market became crowded in the early years of the 1970s, manufacturers looked to provide a new type of electronic calculator – the electronic slide rule. Slide rules had been used by generations of students, scientists, mathematicians and engineers where trigonometric and logarithmic calculations were required. The drawbacks of the slide rule were that a person had to learn how to use one, and they also had relatively poor accuracy. An electronic version which gave a result at a press of a button and with great accuracy would therefore be a very desirable tool. They became known as 'scientific' calculators.

The electronics required to perform these maths functions – a mixture of arithmetic chips and ROM memory chips – together with better keyboard technology and displays, were all becoming available in 1972. They also moved to what is known as a 'scientific' display, where a number is split into its 'mantissa' or decimal part, and the 'exponent', which is the power of ten the mantissa is multiplied by. So, for example, a scientific calculator displays 1.0 10^3 instead of 0.001.

This chapter describes ten of the most important scientific calculators on the journey of increasing complexity in pocket-sized calculating devices.

Hewlett Packard HP-35

Date: February 1972
Distinction: First handheld scientific calculator
The HP-35 holds a unique place in calculator evolution. When Hewlett Packard released it in 1972, it was the world's first handheld scientific calculator. It could perform trigonometric, logarithmic and exponential calculations, and so effectively made the slide rule, which had been used for generations, obsolete. It went on to sell in the hundreds of thousands.

The Hewlett-Packard Company was founded in a garage in Palo Alto by Bill Hewlett and David Packard in 1939, where they produced electronic test and measurement equipment. HP's place in US technological history was firmly cemented when their original garage at No. 367 Addison Avenue was marked with a plaque calling it the 'Birthplace of

The first electronic slide rule – the HP-35.

Silicon Valley'. Hewlett and Packard formally established the Hewlett-Packard Company on 2 July 1939 after they won a contract to provide equipment for Walt Disney's production of *Fantasia*. This was a first step in growing into a multinational corporation, and by 2007 they had become the world's leading PC manufacturer. When HP introduced their first calculator back in 1972, it was the first in a line of high-quality, high-specification machines that would build the reputation of the company.

As the HP-35 was the only calculator they made at the time, it wasn't marked with any model number.

Named for its thirty-five keys, it offered the trig functions: sine, cosine tangent and their inverses via the ARC key, for angles in degrees. Also provided were common and natural logarithms and their inverses. To perform these functions, the HP-35 contained a total of five ICs: an MK6020 Arithmetic & Register chip, an MK6021 Control and Timing Chip, and three ROM chips – the MK6022, MK6023 and MK6024. The log and trig functions are calculated from mathematical series and the HP-35 and subsequent models use a method called the 'Cordic technique' to efficiently calculate the answer by a process of iteration. There is no [=] button as Reverse Polish Notation (RPN) is used where the two operands are entered first followed by the operator. So, for example, to add 2 and 3 together you would type [2] [ENTER↓] [3] [+] to get the answer 5. The result is displayed in ten digits for the mantissa, or decimal part, and two digits for the exponent part. Each LED digit has a lens over it to magnify the small numbers. Physically it measures approximately 3 inches × 6 inches × 1.5 inches, and requires three AA-size rechargeable cells. These last around three hours, but the calculator can be used while the charger is plugged in.

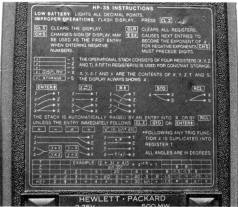

The display has ten digits of precision and instructions are on the rear.

There is a constant memory accessed by the [STO] and [RCL] buttons plus four other memories or registers (X, Y, Z and T). These are arranged as a stack, and so any entry on the display goes into X, the running total is held in Y and trig results go into Z and T. The stack can be manipulated by three buttons. The [X↔Y] button exchanges the contents of X and Y, the [ENTER↓] button pushes the values down in the stack, so X goes into Y, Y goes into Z etc and the [R↓] button does the reverse, i.e. T goes into Z, Z goes into Y register, etc.

Initially, a small number of HP-35s were produced for the engineers at HP. At the time, the company really did not see a market for such a device, but when it did go on sale for $395 (about £165), and despite its high cost, HP went on to sell more than 300,000 units. HP had a long history of making quality electronic instruments and this, the first in a range of professional calculators, was well made and reliable.

The one photographed is an early model and is unusual in having no serial number on the back. Early models have a raised dot on the '5' button and a bug where if you enter 2.02 [ln] [e^x] the result should be 2.02 but in fact you get 2.00. Later models differ slightly in that they are labelled 'Hewlett•Packard 35', and have a single silver trim bar below the display. Also, function names are printed on the buttons themselves rather than the casing, and the buttons are rectangular not square.

In 2009 Hewlett Packard was awarded the prestigious 'IEEE Milestone in Electrical Engineering and Computing' for the HP-35 calculator, as the first handheld scientific calculator. Hewlett-Packard was the first but the competition was not far behind. Hewlett-Packard continue making calculators today.

Hewlett Packard HP-80

Date: February 1973
Distinction: First handheld calculator with business and finance functions
Hewlett-Packard's second calculator, the HP-80, was an extremely considered purchase in 1973, because the $395 purchase price equates to £2,000 in today's money. But the HP-80

wasn't aimed at the man or woman in the street, rather it was for people who worked in the financial sector such as accountants, bankers and mortgage advisers. Prior to calculators, financial calculations were worked out from tables of numbers to give approximate answers which were then refined using a slide rule or pen and paper to get a final answer. Even then, the accuracy was limited and prone to error. The HP-80 replaced most common financial tables with a set of over thirty built-in programs to work out financial calculations with a few button clicks. According to William R. Hewlett, president of the company: 'The HP-80 can perform virtually all calculations involving the relationship between money and time quickly and easily ... It can be operated for five hours from nickel cadmium batteries.'

The hardware of the HP-80 closely resembled the HP-35, but whereas the HP-35 had three ROM memory chips, the HP-80 had seven to store 18,000 bits of information for table data and code to run the programs. The programs were actually solving equations whereas the HP-35 would solve functions. The designers of the calculator used the Newton-Rhapson method, where the input is changed slightly and the resulting output compared to what is required. This is repeated over and over until a solution is found, and is known as iteration. There is also a built-in 200-year calendar that can be used to find the number of days between any two dates, or find the day of a past or future date. In terms of size, weight and batteries, the HP-80 matches the HP-35, and the display is the same red LED ten plus two digits. To protect the owner's investment, the calculator came with both hard and soft carry cases, instructions, a quick-start guide and battery charger.

The HP-80 has forty buttons, including a gold-coloured 'Shift' button for use with those that have two functions. Differences with the HP-35 were a lack of trig and log functions, and the [xy] and [ENTER] buttons were re-labelled. Explanations of all the functions would require a lot of space but what follows is an example problem that the calculator can solve.

Hewlett-Packard's business calculator – the HP-80.

The expensive HP-80 came with a sturdy case.

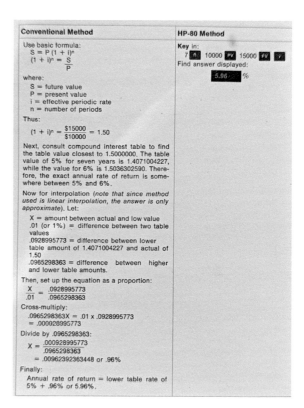

Conventional Method	HP-80 Method
Use basic formula: $S = P (1 + i)^n$ $(1 + i)^n = \dfrac{S}{P}$ where: S = future value P = present value i = effective periodic rate n = number of periods Thus: $(1 + i)^n = \dfrac{\$15000}{\$10000} = 1.50$ Next, consult compound interest table to find the table value closest to 1.5000000. The table value of 5% for seven years is 1.4071004227, while the value for 6% is 1.5036302590. Therefore, the exact annual rate of return is somewhere between 5% and 6%. Now for interpolation (*note that since method used is linear interpolation, the answer is only approximate*). Let: X = amount between actual and low value .01 (or 1%) = difference between two table values .0928995773 = difference between lower table amount of 1.4071004227 and actual of 1.50 .0965298363 = difference between higher and lower table amounts. Then, set up the equation as a proportion: $\dfrac{X}{.01} = \dfrac{.0928995773}{.0965298363}$ Cross-multiply: .0965298363X = .01 x .0928995773 = .000028995773 Divide by .0965298363: $X = \dfrac{.000028995773}{.0965298363}$ = .00962392363448 or .96% Finally: Annual rate of return = lower table rate of 5% + .96% or 5.96%.	Key in: 7 n 10000 PV 15000 FV i Find answer displayed: 5.96 %

A page from the manual showing how to calculate compound interest.

'If you buy a house today for £100,000 and expect to sell it in seven years' time for £150,000, what would be the annual rate of return on your initial investment?' To work out the answer, key in 7 (the number of years) and press [n]. Key in 100000 (initial investment) and press [PV] (present value). Key in 150000 (amount to be returned) and press [FV] (future value). Then press [i] (interest). The answer displayed is 5.96 per cent. For more accuracy, press the gold button to give 5.963402. To do the same without the calculator, you would have to use tables to get an approximate answer of between 5 per cent and 6 per cent, and then use a method called linear interpolation which involves several more steps of multiplication and division to get a final answer.

The HP-80 was successful in the financial world and was in production until 1978.

Casio fx-10

Date: August 1974
Distinction: Casio's first scientific calculator
For schoolchildren in the 1970s and 1980s, if they used a pocket calculator in a maths lesson, it was probably a Casio. Casio produced a whole range of scientific calculators that were small, well made and affordable. Model numbers prefixed by 'fx' are their scientific models and in the decades from the 1970s to the present day Casio added dozens to the range. The first 'fx' though was the 'fx-10' in 1974, the 10 denoting that the calculator had ten scientific functions, and the prefix 'fx' was chosen because f(x) is the mathematical notation to represent a function of x.

A ten-function calculator from Casio.

The fx-10 displaying the number 1020304.5.

Measuring 3.75 inches × 5.9 inches × 1.3 inches, it is a fairly chunky handheld machine, powered by four AA-size batteries or an AC adapter. Inside, the display is composed of a single Digitron vacuum fluorescent display tube giving eight digits with floating point, or seven digits if the displayed number is negative. There is no exponent display and so overflows (showing all 0s) if the result is greater than 99,999,999. Note that 0s are displayed half-height as was common at this time but looks odd today. On the top row, the functions are: common logarithms [log] and natural logarithms [ln]. Next to those are the exponent function [e^x], [a^n] which raises a number to a power of n, and finally the reciprocal function [1/x]. On the bottom row is a sexagesimal button [° " '], which takes an angle in degrees, minutes and seconds and converts to decimal for use with a trig function. Next to this are the sine, cosine and tangent buttons and finally square root [√]. Note that there is no button to calculate an inverse trig function on this model, but the manual gives the formulas to calculate them by hand. At just under $100 when it was released in the USA, it was only a quarter the price of the HP-35 released two years earlier and sold well.

Sinclair Scientific

Date: 1974
Features: Small, low-cost scientific calculator
After the success of the basic Cambridge range, Sinclair's next challenge was to produce a scientific calculator that could perform trigonometry and logarithms. Of course, it had to be both smaller and cheaper than anything else on the market and at £49 plus VAT, this was achieved with the 'Sinclair Scientific'. The big challenge was to cram all of the new functions into a single chip so that the existing Cambridge style case could be used and to keep the costs down. The problem, however, was that developing a brand-new calculator-on-a-chip was expensive, time-consuming and risky. In typical Sinclair style, he found an innovative alternative solution.

The pocket-sized Sinclair Scientific.

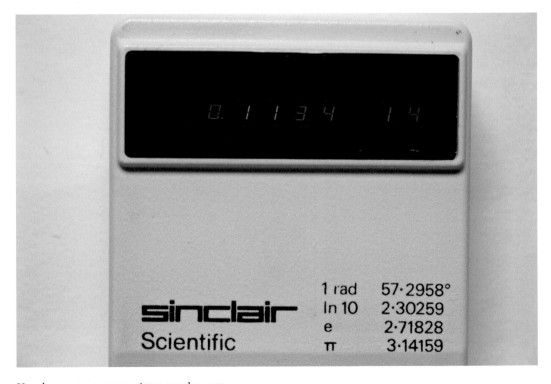

Handy constants are written on the case.

At the time, Texas Instruments offered basic off-the-shelf calculator chips such as the TMS0805. However, these were not just 'dumb' logic chips, but were programmable microcontrollers with built in RAM and ROM, i.e. simple computers. One of Sinclair's engineers, Nigel Serle, used great ingenuity to re-program the TMS0805 chip, and into it was squeezed all the functions needed for his scientific calculator. This was not an easy task by any standards. Although the resulting calculator was not as accurate as other machines of the time such as the HP-35, Sinclair's offering was much cheaper and filled a niche market. As well as offering the new trigonometric and logarithmic functions, the Scientific differed from the Cambridge range in two ways. Firstly, numbers were entered and results displayed in scientific format with a five-digit mantissa and two-digit exponent. For example, as there was no decimal point button, 1,234,500,000 is keyed in as 1 2 3 4 5 [EE] 0 9 and displayed as '1.2345 09'. Secondly, the calculator uses Reverse Polish Notation (RPN), so to calculate: '3 + 4', the following button presses are required: 3 [+] 4 [+]. Here, confusingly, the first plus is just to separate the two numbers and not to add them. The second plus actually adds the numbers and gives the result of 7 straight away, no 'equals' required. Using RPN was a side effect of having to customise the chip in the way they did and RPN was a more efficient way of doing things with limited space inside the chip. Also, it saved space on the cramped keyboard by needing one button less. It certainly took some getting used to if you had previously owned a standard Cambridge calculator.

A couple of years later in 1976, Sinclair launched a more conventional scientific calculator, the Cambridge Scientific, which did away with RPN and exponents and went back to an eight-digit display. As well as having log and trig functions, it had a memory, square root, reciprocal and the ability to switch between degrees and radians.

Texas Instruments TI-30

Date: 1977
Features: Stylish, low cost
Texas Instruments were a prolific and successful manufacturer of handheld calculators in the 1970s and the TI-30 was among the most popular of all LED calculators. It was initially priced around £14 and was manufactured into the 1980s. They can still be found on internet auction sites for just a few pounds. Measuring 4.0 inches × 6.25 inches × 1.25 inches, it is quite large, but Texas Instruments thought it was cheaper to retain the case they had been using for a few years rather than design a new one. Inside, there was just a single integrated circuit – a TMC0981 – connected to the nine-digit LED display module and space for a 9V battery.

The keyboard looks quite stylish with its metal brushed-gold effect and the functions offered were standard for the day. Along with common and natural logs are the three trig functions and their inverses. The [DRG] button allows for angles to be input in not only degrees but also radians and grads. The chosen units are indicated by either a single or double quotation mark at the beginning of the display. There is a [y^x] button to raise a number y to the power x, a memory, a constant and brackets. Also, a useful [π] button, plus square, square root and reciprocal functions. Interestingly, trig functions take a relatively long time to calculate (a second or so), and before the answer is displayed, the first digit

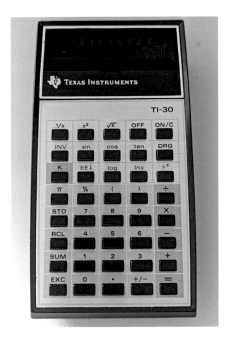
Texas Instrument's popular calculator.

animates in a pattern which may be intentional or just the display going a bit random while it is calculating. Texas Instruments is one of the few early manufacturers which is still in the calculator market today.

Commodore S61

Date: 1977
Distinction: Specialist statistical calculator
Commodore was one of the main calculator designers in the 1970s and the culmination of their scientific range was the S61, known as the 'Statistician' after its comprehensive list of statistical functions. It was therefore aimed at students, scientists and professionals that worked with large amounts of statistical data. Commodore referred to the S61 as 'third generation scientific' in that it had pre-programmed functions. Even if not everyone agreed with this categorisation, the S61 was impressive with its sixty buttons, many of them dual-use. Commodore also produced a navigation calculator with sixty buttons called the N60, so why the Statistician is named 'S61' and not 'S60' is a mystery!

Commodore must hold the record for the widest pocket calculator when they designed the S61 as it measures 4 inches across – twice that of the Sinclair Scientific. The display is ten digits plus two more for the exponent. Along with the statistics functions, there are eight separate memory registers and a random number generator. Statistics is a specialised branch of mathematics and so most people would not recognise all the functions that the S61 could solve. Many people who took a GCSE in maths though would probably still remember the basics such as mean, median, mode and standard deviation. To calculate the standard deviation of a group of numbers, you enter the first number and then press

The S61 and its impressive number of buttons.

the $[\Sigma]$ button (top right, below the on/off switch). Repeat this sequence for the rest of the numbers. To find the average (mean) of the numbers, press [x-bar] (two buttons to the left). Pressing the function button [F] then [x-bar] calculates the standard deviation of the group of numbers entered. Others may be familiar with linear regression, Poisson distribution and permutations and combinations. Either way, the comprehensive 111-page user manual gave plenty of examples of how to use all the different capabilities of the S61.

Casio fx-19

Date: 1976

The Casio 'fx' range of calculators, familiar to most schoolchildren in the 1970s and 1980s, was both iconic and practical. They were well designed and built to last, but were also packed with features. The models in the range had a similar look and feel, with their metal-covered keypad and array of grey rectangular functions buttons above the black number buttons, so that if you upgraded to a better model, it would instantly feel familiar. The range is still made today, with well over 100 models produced so far. Beginning with the fx-10, which had ten function buttons, the fx-19 in 1976 had, predictably, nineteen. New additions were inverses of the three trig functions, plus radians and gradians units for angles. There were five new statistics buttons including mean value and standard deviation of a set of numbers, and fractions could be input directly. There was also a memory and display of angles in degrees, minutes and seconds.

A well-used fx-19 displaying the fraction 1¼.

Arithmetic with fractions was a new feature that Casio introduced and kept on most future models. The fraction button (top right) was used to separate the numerator, denominator and any whole numbers. Then addition or subtraction could be performed with another fraction. The scientific functions are mostly one function per button, i.e. no shift button required. The statistical functions are slightly different in that when the selector switch under the display is set to 'SD', the five buttons on the bottom row with brown legends are used for entering data and calculating the mean, etc. The DMS button [° ' "] displays a decimal angle in sexagesimal (base 60) with the degree symbol used to separate degrees, minutes and seconds on the display. This would become a standard feature on all Casio scientific calculators.

Casio fx-31

Date: 1978

Two years after the fx-19, Casio launched the fx-31. There were only minor differences. Physically, the fx-31 is slightly smaller and thinner and requires only two AA-size batteries instead of four. The buttons are slightly smaller, and the on/off switch is now on the side. Most of the function buttons (reduced to fifteen from nineteen) double up with their inverse function, e.g. square and square root. The [INV] button is pressed first to access the

An fx-31 displaying the angle 30 degrees, 39 minutes and 28 seconds.

function coloured in brown. Other new features were six levels of parentheses and a handy button to display the constant 'pi'. One omission was the fractions button. The blue-green vacuum fluorescent display has eight digits or six + two when in scientific notation. Due to their relatively high power consumption, VFDs were coming to the end of their life in the later 1970s and would shortly be replaced by LCD displays. Many Casio calculators at this time were re-badged and sold by the Boots company of Nottingham. The fx-31 was sold under the name 'Boots 425 Scientific'.

Casio fx-350

Date: 1979

By the late 1970s, pocket calculators had moved on from the blue-green glow of the fluorescent tube display to the black-on-white of LCDs. The power consumption was so much lower that the large cylindrical AA-size batteries could be replaced with a single 3V button or coin cell. These tiny batteries would give months of typical use. The fx-350

An fx-350 in its handy wallet.

model, introduced around 1979 carried on the same design style of earlier models such as the fx-31, but the metal casing now covered the whole front face. Size and weight had been reduced considerably, with the thickness now being only 8 mm – true pocket size – and the number of functions offered increased to fifty-four. Some of the new ones were co-ordinate conversion between rectangular and polar, a random number generator and a couple of functions called FIX, which specifies the number of places after the decimal point, and SCI (Scientific Notation). SCI lets you set the number of digits for the mantissa, for example when the speed of light in m/s is displayed (299792458), selecting SCI and pressing [3] displays as three figures, i.e. '3.00 08'.

Casio fx-451

Date: 1980
Feature: Foldable calculator
The fx-451, from around 1980, was the ultimate in pocket scientific calculators. It had everything – slim, lightweight, solar-powered and packed full of features. There were too many buttons to fit under the display alone and so a second keypad was added to the

Solar-powered fx-451 has buttons built into its wallet.

right-hand-side and hinged so that it could be folded in half. The keypad is touch-sensitive and similar to the keyboard of a Sinclair ZX80 computer of the same period. When folded up, it measures only 4.7 inches × 3.1 inches × 0.4 inches thick.

The list of functions available is truly impressive. As well as all the functions of the fx-350, the fx-451 offers functions for computer programmers such as conversion between number bases (binary, octal and hexadecimal), and six logical operators including OR, AND, NEG. For physics students there is an array of physical constants stored in the calculator, such as 'c', the speed of light, accessed by the shift key and one of the number keys. For general users, there are eight unit conversions in yellow on the keypad for such things as temperatures and weights. Pressing the mode button followed by the mode number (written below the display) selects one of the ten modes available, such as statistics (SD) or fixed decimal point (FIX), and the mode selected is indicated on the display. One issue with the innovative design was that opening and closing the wallet stressed the flexible wires in the spine and they could break, rendering the keypad unusable. But it could be forgiven for that because while it worked it was awesome!

3

Programmable Calculators

By 1974, scientific calculators were available that provided most of the mathematical functions that were needed by students, engineers and scientists. But if you needed to perform a complex calculation over and over again using a conventional calculator, it would soon become tedious. For instance, to plot a graph of '$y = x^3 + x^2 + x$', for each point you would have to compute 'x cubed' and then 'x squared' and then add them both to 'x' to get a value for 'y'. This is time-consuming if you have more than just a few points on your graph. An easier way is to automate the process, and this is where programmable calculators come in. A small program is written to take a value for 'x', find the cube and the square and add the three terms to give the result for 'y'. Of course, this is just a very basic example and programs with dozens or even hundreds of steps became possible.

Programable calculators are simple computers, but are, of course, limited by the size and sophistication of their programs. This chapter describes six of the most important programmable calculators of the period.

Hewlett Packard HP-65

Date: January 1974
Distinction: First handheld programmable calculator
Hewlett Packard had a history of calculator innovation, producing the first scientific calculator, the HP-35, in 1972 and the first business and finance calculator, the HP-80, in 1973. A year later they produced the first programmable calculator, the HP-65. It was advertised as combination calculator and personal computer that you could carry around with you. It is the same size as the other handheld calculators in the family and also runs on batteries or from a charger.

The HP-65 is a comprehensive scientific calculator with fifty-one functions and a display of ten + two digits in scientific notation. Its programming abilities are equally impressive. Not only can it run a 100-step program, but the programs can be stored on magnetic cards via a tiny built-in magnetic card reader. HP offered hundreds of pre-programmed cards for maths, statistics, surveying and even medical programs. Magnetic cards were not new

The programmable HP-65.

but had previously only been used with desktop computers. They were useful because, as the memory was volatile (it lost its contents after power was removed), without them, the program would have to be typed in again by hand after the calculator's power was turned back on.

When power is first applied to the HP-65 the top row buttons (A, B, C, D, E) are defined as functions [1/x], [√x], [y^x], [x-y], and [R↓] (roll down the operational stack). However, the functions of these buttons can be changed, by loading a program from a previously recorded magnetic card via the card reader slot. The new functions of the buttons can be written on the card and the card inserted into the window above the top row of buttons to show the new functions. For example, the standard program pack supplied a compound interest program, which changed these buttons to [n] (number of periods), [i] (interest rate), [PV] (present value), [FV] (future value), and [COMPUTE], similar to buttons on the HP-80 financial calculator.

This was a big leap forward in 1974 in bringing affordable computing power to the masses. At $795 it was mostly only available to professionals who needed it for their work, but that would soon change. If you can find a HP-65 for sale today, prices start at around £300.

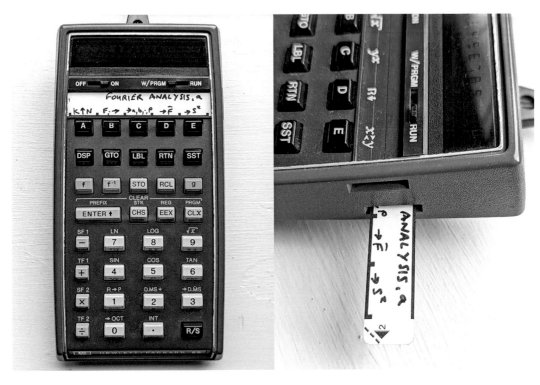

Magnetic cards were used to load a program.

Cambridge Programmable

Date: September 1976
Features: Low-cost, thirty-six-step program calculator

On 1 September 1976, the Cambridge Programmable was released (marketed in the United Sates as the Radio Shack EC-4001) at a cost of $34.95. It took the larger 9V PP3 battery, which resulted in a bulge at the rear. From a practical point of view, the bulge meant that it wouldn't sit flat on a desk and so to remedy this, Sinclair supplied a plastic tray for the calculator to sit in.

Being so small and therefore having few buttons, most had two or three uses. To access the functions marked above the buttons, shift (two triangles) is pressed once and an 'F' appears on the display to indicate upper case mode. Pressing the shift button twice accesses the lower functions indicated by a 'G' on the display. As well as having the scientific functions of the Cambridge Scientific, it has an eighteen-byte memory allowing a thirty-six-step program to be entered. This was a small size in comparison with other programmable calculators of the time such as the HP-65, which had 100 steps. However, Sinclair went to great lengths to show what could still be achieved in thirty-six steps by providing four volumes of ready-made programs to type in – Finance and Statistics, Maths, Physics and Engineering and Electronics. This amounted to nearly 300 programs covering such things as temperature conversions, solving mathematical and physics equations, mortgage calculations and even a moon landing game (topical in the 1970s!).

Sinclair's tiny programmable calculator.

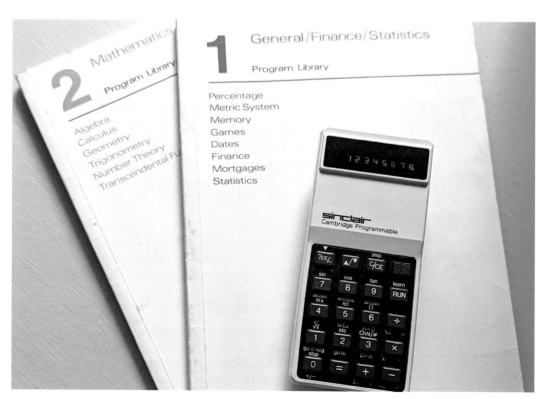

Several books of programs were available.

In the game, the player had to enter a time to 'burn' or 'coast' the rocket engine and then the program would say how much fuel was left, the height above the surface and the vertical speed. The object of the game was to get the height to zero and the speed to zero before the fuel ran out. Whether the program qualified as a game or not is up for debate, but this was 1976!

A not-so-glowing review in *Which?* magazine from September 1977 summed up the calculator as 'Remarkable for its price. Very portable, but hard to use, not very versatile and very inaccurate. Expensive on batteries. Improved version promised.'

Texas Instruments TI-58

Date: 1977
Features: Programmable calculator with pre-programmed functions
The TI-58 from Texas Instruments is similar in many ways to the earlier Hewlett-Packard HP-65. The TI-58's trump card, however, is that it came with a library of programs already built into a chip fitted into the rear of the calculator. This was much more convenient than loading the programs from magnetic cards. TI referred to the library as Solid State Software, and it could be used straight off, without knowing anything about programming. The library offered such general-purpose functions as unit conversions, complex number arithmetic, matrices, polynomial evaluation and also a diagnostic tool and a number guessing game. Each program can be called up using the [Pgm] button followed by the program number, and used according to the instructions in the manual. The library modules were interchangeable and other more specialist libraries of programs could be purchased, with titles such as Electrical Engineering, Real Estate/Investment and Securities Analysis. This flexibility was the cheapest way to run many different types of program – buy the calculator and then just swap over the libraries.

Some companies used the TI-58 for specialist purposes by producing their own libraries and keyboard templates. One such custom library was for the Harrier jet, a vertical take-off and landing aircraft used by the UK and the US at the time. A navigator could take the calculator with them on the aircraft and calculate flight data such as speed, range and endurance during a mission. Programs could be downloaded from the library into the calculator's memory for examination and diagnostics if there was sufficient space.

Programs could also still be entered manually if required. Pressing the Learn button [LRN] and then a calculation sequence, such as that to convert Celsius to Fahrenheit, causes the calculator to remember the sequence. To run the conversion program, a Celsius value is entered, then the reset and run/stop buttons pressed ([RST] [R/S]) to obtain the answer. The TI-58 has room for a program of 480 steps, with up to sixty registers available for data storage. The more expensive TI-59 model has double the program storage and up to 100 registers. The TI-59 can also use magnetic cards for reading and writing programs. Otherwise, the 58 and 59 are functionally identical. There was also a printer available to print out a program, or its results.

A review in *Which?* magazine from September 1977 said of the TI-58: 'Rechargeable. Extremely versatile and accurate. Excellent facilities; also good as statistical or financial calculator. Liked by our users.'

Texas Instrument's programmable TI-58.

TI-58's innovation was its slot-in
program libraries.

Casio fx-501P / fx-502P

Date: 1979

Distinction: First LCD programmable calculator

The fx-501P and its bigger brother the fx-502P were Casio's offering into the programmable calculator market of the late 1970s. The main advantage over its rivals at the time was that it had an LCD display, which meant low power consumption and long battery life. Being an early LCD display, it required the yellow filter to protect against UV light. It is a very slim calculator and at just 9.6 mm thick would easily fit into an inside pocket, unlike models from HP and TI at the time. Both the 501P and the 502P are fully fledged scientific calculators with fifty-one functions available, and the display is ten digits + two exponent. For programming, the fx-501P offers the full range of options that you would have expected at the time, namely, conditional branching (with up to ten labels to jump to), unconditional branching (goto), subroutines and indirect addressing for both memory access and jumps. These features allow for efficient programs to be written into the 128 steps available (or 256 steps on the fx-502P).

The fx-501P/502P have a standard seven-segment display which means that they can only display the digits 0 to 9 and the letters C, E, F and P. Therefore, program steps that were entered (which may have contained non-displayable characters, like MR (Memory Recall) were converted into two-digit codes when stored in the calculator. Although programs would be lost when the batteries ran out, a cassette interface could be purchased

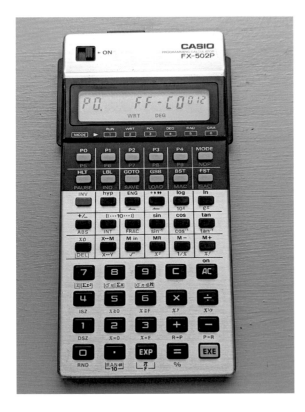

An fx-502P in program mode.

A large number of programs were available.

The FA-1 cassette recorder adapter. Note the keypad overlay for music programs.

called the FA-1, which enabled programs to be written to cassette tape. The cassette adapter takes the form of a tray that the calculator sits in and connects via a seven-pin connector at the top of the calculator. An impressive 225-page book of programs were supplied covering areas such as maths, electrical engineering, mechanics, physics, statistics, medicine, surveying, banking, navigation, games and even music. In the music programs, each note was represented by a code, and the tempo was specified by a value in a register. The music could be recorded onto tape and played back via the cassette recorder's earpiece or speaker. An early synthesiser!

Hewlett Packard HP-41C

Date: 1979

Feature: Connection to peripherals to form a 'system'

When HP brought out the HP-41C in 1979, it was their most powerful calculator to date. Instead of being just a standalone calculator, it could be connected up to other devices via its system bus, just like a real computer. The four input/output ports on top of the calculator allowed for the insertion of either modules for expansion of RAM (user memory) or ROM (for application modules) or the connection of peripherals such as a printer and a magnetic card reader.

Another major innovation of the HP-41C was the alphanumeric display. This was HP's first calculator with an LCD display and the first to display both letters and numbers.

The HP-41 can display the whole alphabet.

Top view showing two expansion modules fitted and space for two more.

This was a big leap forward in making programming a more pleasurable experience. Alphabetic characters allowed for the naming and labelling of programs and functions, prompting for data with meaningful words and display of useful error messages. To enter text from the keyboard, the [ALPHA] button is pressed and then a string could be entered of up to twenty-four characters including punctuation marks. Each button produces the letter labelled in blue on its front. The text goes into the 'ALPHA Register' where it can be used in a program.

The HP-41C offers over 130 functions but has only sixty-eight available on the keypad. To access the rest, a function is assigned to the button of your choice as required. For example, it makes sense to assign the command 'PRINT' to a button only when the printer is attached. Pressing any function button down and holding it displays the name of the function for approximately one-half second. When the button is released, the function is executed. Buttons that are reassigned to a different function are maintained while the calculator is switched off. The programming capabilities of the HP-41C are comprehensive with branches, loops, subroutines and indirect operations available for advanced program writing. Program editing and debugging are straightforward, and the user manual was well written. Two other models were made: the HP-41CV, with five times the memory, and the HP-41CX, with extended functions and time calculations was introduced in 1983. The HP-41 series were a great success for HP with at least 1.5 million units sold.

Casio fx-7000G

Date: 1985

Distinction: First graphing calculator

As well as being programmable, the fx-7000G has the accolade of being the world's first graphical calculator. The LCD display is capable of displaying eight lines of sixteen characters or a 64 × 96 pixel graph. Not only does it have the most comprehensive set of mathematical and statistical functions of any scientific calculator, it can store up to ten programs, with a maximum of 422 steps in total. Having a screen means that it is larger than your average LCD calculator at 3.3 inches × 6.5 inches.

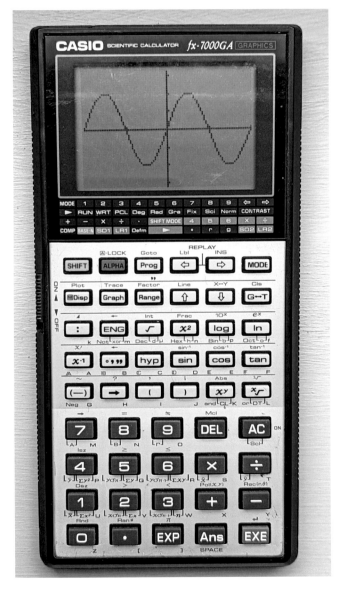

The fx-7000G displaying a sinusoidal function.

Graphs are a useful way of visualising a mathematical formula, and having twenty built-in plots made this very easy. For example, to see what a sine curve looks like requires the following five-button sequence: [MODE] [+] [Graph] [Sine] [EXE]. After the graph appears, the Trace function can be turned on ([SHIFT] [Trace]), which places a blinking cursor onto the graph and it can be moved around using the right-arrow key. The value at the cursor is displayed below the plot. The plots are nicely scaled to fit on the screen and afterwards the ranges of the axis can be changed as required using the [Range] button. Up to two different plots can be displayed at once and just about any formula or built-in maths function can be drawn with a little effort in reading the comprehensive user manual.

Although it is called a scientific calculator, with the full set of trig, log, statistics and number base functions, it also has quite sophisticated programming abilities. Formulas can be incorporated into programs so that they can be repeatedly executed, with unconditional jumps or conditional jumps depending on values either input or calculated in the program. Programs automate a sequence of steps, saving time and effort. When running, a typical program might ask the user to enter a series of numbers, such as data points or constants from an equation, then compute and display the answer. The user manual listed fourteen such programs to type in; for example, working out the surface area and volume of an octahedron. However, the program memory is volatile, i.e. it is lost when the batteries run out. So even though the three 3V button cells last around 100 hours, any programs that have not been saved prior to a change of batteries have to be re-entered. The fx-7000G was a popular calculator and sold well, and second-hand examples can be bought fairly cheaply.

Different screen displays. The centre photo is a statistical plot.

4

Pocket Computers

It can be argued that programmable calculators marked the end in pocket calculator development, and that they did everything that a user wanted from a calculating machine. Some of the companies that made electronic calculators, such as Texas Instruments, Commodore and Sinclair, went on to produce successful home computers in the 1980s. These were generally desktop machines with full-size keyboards and connected to a television for their display. But in-between pocket calculators and home computers, a few companies produced pocket computers. These were handheld devices, still with a single-line display like pocket calculators, but with the addition of a small QWERTY keyboard. They were a sort of supercharged programmable calculator but with a proper programming language, unlike their predecessors. A couple of examples are described here to show how far pocket-sized electronics had come in a decade.

Sharp PC-1211

Date: 1980
Distinction: First pocket computer
Before the home computer boom of the 1980s when most homes would have a Sinclair Spectrum or a ZX80 to plug into their TV, Sharp brought a pocket version to market in the PC-121x series. The first of these was the PC-1210 with 896 bytes of memory but a year later, the PC-1211 and PC-1212 had double that. The PC-1211 was also marketed by Radio Shack/Tandy as the TRS80 PC-1. They measure 6.9 inches × 2.75 inches × 0.7 inches and feature both a QWERTY keyboard and numeric keypad, a twenty-four-character 7 × 5 dot-matrix LCD display and weigh just 170g. The processing power is provided by twin four-bit CMOS processors and their low power consumption gave 300 hours of use from four button cell batteries. Although it could not be connected to a TV, there are connections for a tape recorder for program saving, and a printer.

The PC-121x uses the 'high-level' language BASIC with its easy-to-learn English-style syntax for commands and statements. BASIC has a history stretching back to the 1960s but from the late 1970s, it became the de-facto language on almost all computers

Sharp's compact computer, the PC-1211.

The printer and cassette interface were a useful accessory.

due to its ease of use. Computers at the time had tiny amounts of memory and so manufacturers had to create simplified versions for their machines. Even so, the PC-1211 offers a more-than-adequate version, with control statements such as 'FOR..NEXT..STEP', subroutines via the GOSUB command and multiple statements per line.

There are also fifteen mathematical functions, and it can be used as an ordinary scientific calculator if required with a ten + two digit scientific display. But for programming, their small physical size means that typing is awkward, and the twenty-four-character display means that a line of code longer than that has to be scrolled to view it all. They were therefore of limited appeal to the general public and only sold to a niche market of scientists and engineers. Even so, the PC-121x series were triumphs of miniaturisation. They are hard to find today and after forty years, any seen for sale may have faulty screens like this one has.

Sharp PC-1500

Date: 1981 (Japan), 1982 (Europe)
Feature: 8-Bit pocket computer
After the successful PC-121x series, the PC-1500 was the second pocket computer released by Sharp. There are lots of minor differences, perhaps the most significant was being based on a 'proper' 8-bit processor running at 1.3MHz. Physically it is a bit larger at 7.4 inches × 3.3 inches and bulkier at 1 inch thick. This is to accommodate the four AA-size batteries, but on the plus side the batteries were cheaper and easier to replace than the previous button cells.

The display increased from twenty-four to twenty-six characters and gaps between characters were removed giving 7 × 156 pixels for simple graphics. An increase in memory to 3.5KB gives nearly 2K (1,850 bytes) available for programming and a further 512 bytes for fixed variables and the display buffer. But the big advantage over its predecessors was its expandability. There were RAM modules up to 16KB, a docking station for cassette interface and a colour plotter, and an external board with various interfaces. The version of BASIC is similar to that in the PC-121x machines but the PC-1500 can also be programmed in machine code. So, the whole package added up to being a useful machine popular amongst hobbyists and also used by several companies for their staff in the field.

Two years later there was a PC-1500A released with around 6KB of memory available for programs, and as with the PC-1211, there was a custom manufactured version of the PC-1500 built by Sharp for Tandy Radio Shack, called the TRS-80 PC-2.

After pocket computers, handheld electronic devices continued to evolve. Next came PDAs (personal digital assistants) such as the Psion Organiser, which had an address book and electronic diary and competed with the paper Filofax. More sophisticated PDAs came next such as the powerful Psion Series 3 with a database, spreadsheet and word processor and even internet connectivity. Ultimately, of course, all handheld devices would be replaced by the smartphone and tablet.

However, pocket calculators are still around and can be found on the shelves of most stationers for just a few pounds. Even though your smartphone will have a calculator app, there is no substitute to pressing the real buttons of a pocket calculator when it comes to doing the home accounts, or seeing what upside-down words you can spell during a maths lesson.

The compact PC-1500A.

5

Thinking Outside the Box

Marketing departments of big companies are tasked with the job of seeking out ways to sell the consumer something they already have, or maybe something they don't need, or don't think they need. This is a tricky job but one way is to repackage the product. Another is to combine it with something else so that the consumer feels they are getting something for nothing. In this way, the electronic calculator found new life inside packaging that wasn't just the boring old square-box shape. The miniaturisation of the calculator meant that from the mid-1970s onwards, it was possible, just, to incorporate a calculator into some other device such as a pen or a watch. These new consumer products were never going to have mass-market appeal and make the companies behind them a fortune, but they were

Not all calculators look the same.

a demonstration of what was possible, and for manufacturers to show off their technical skills. This chapter looks at some different applications for the electronic calculator.

Hosiden 'Calcu-Pen'

Distinction: First calculator in a ball-point pen

The 'Calcu-Pen', as it was known, was released in Japan and the USA around 1975, and it was a demonstration of how far electronics had been miniaturised by the mid-1970s. Although the pen was a quite a bit thicker than an ordinary ballpoint at 0.6 inches diameter and so awkward to write with, it was a great gadget for its novelty value.

It was made by Japanese company Hosiden Electronics, who specialised in making tiny switches. Each of the five buttons is a four-way switch and so for example, you would press on the upper part of the first button for a '1' and press on the left-hand side for a '3'. This four-in-one switch was a great idea for fitting many buttons into a small space but made calculations error-prone for those without tiny fingers! There are just four functions, plus per cent, and power is from a small cylindrical battery. Unscrewing the end cap reveals the battery compartment, and twisting the end cap switches the calculator on and off. The Calcu-pen was expensive to start with at £45 but two years on it had fallen to around £11. They are quite rare items today.

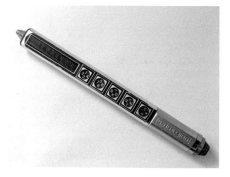

Above: Calculator meets pen – the Calcu-pen.

Right: Not very practical as a pen or calculator.

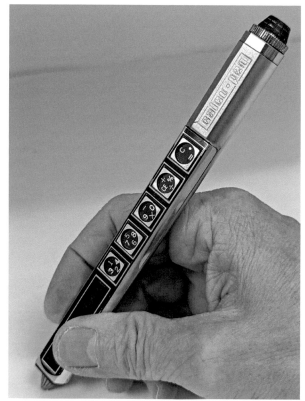

Alco Mickey Math

Feature: Children's calculator

A number of calculators were produced especially for children in the 1970s and this one from Omron was introduced in the US in 1975. Its ergonomic casing with Disney's Mickey Mouse character was designed to make it attractive to a young audience. Being called 'Mickey Math' and not 'Mickey Maths' was a sign that it was made for the US market, but a company called Alco based in Florida distributed it abroad, and some made it to the UK.

It measures 7.5 inches × 7.5 inches × 1 inch and is ruggedly constructed, with a built-in carry handle. Inside, the simple circuit board contains a single chip, the Omron HD3639, a handful of capacitors and resistors and the 'Itron' vacuum display tube. The six-digit display is blue fluorescent, common around this time, and the calculator requires four AA-size batteries. Being a children's calculator, it was kept simple with just the basic four functions and eighteen buttons. Mickey Math came with a storybook-like user guide which attempted to teach children the basics of arithmetic.

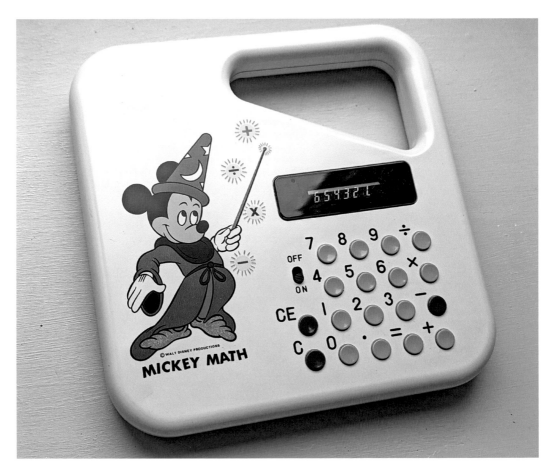

The Mickey Math, a fun calculator for children.

Texas Instruments Little Professor

Distinction: First calculator-based teaching aid

A year after Mickey Math, Texas Instruments produced their own version of a calculator for kids called 'Little Professor'. It was quite an eye-catching design with its bright yellow case and red hat, through which sums could be displayed. It was aimed at five-year-olds and above and the Little Professor's teaching method was to ask questions and mark the child's answers. To start, one of the four levels of difficulty is set using the LEVEL button. Next, the function to be tested is chosen using the SET button followed by +, -, × or ÷. Then the GO button is pressed to display the first calculation, selected from 16,000 pre-programmed

The Little Professor asking the question '2 + 4 = ?'.

A range of game cards were provided.

sums. If the correct answer is entered, the next calculation is displayed. If a wrong answer is entered, 'EEE' is displayed and two more attempts are allowed. After ten questions, the number of correct answers is given. Being a simple game format, it encourages children to try and beat their previous score or beat their friend's scores. It also came with eighteen games to try on printed picture cards, for example a snakes and ladders game where a counter is moved forward for each correct answer.

The Little Professor was popular enough for it to have been produced continuously from 1976 to the present day. There have been many small changes over the years such as an LCD screen in the 1980s and adding a solar panel instead of the 9V battery in the 1990s, but they look basically the same as the original and remain popular.

Casio ST-1

Feature: Calculator/stopwatch combination

To measure lap times in sporting events traditionally required a mechanical, wind-up stopwatch, which looked like an old-fashioned pocket watch. Then, in 1977, the Casio Computer Company introduced this combined calculator/electronic stopwatch, which would time a race to tenth-of-a-second accuracy, and measure lap and split times. The vertical selector button determines which of the four timing modes to use, and in the COMP position (Computation) it is a calculator. It has square root, per cent and also a memory, which can store a time from the watch.

The ST-1 was a well-thought-out design with a rugged aluminium case and a bright fluorescent display which is recessed to be seen more easily in bright sunlight. Other features are a right-hand side button for start and stop and a clear protective cover for the keypad to prevent accidental button pushes whilst timing. It measures

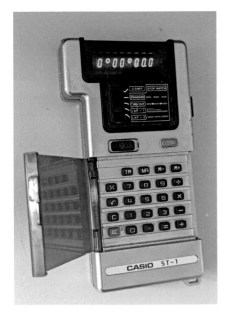

An early electronic stopwatch – the ST-1.

2.9 inches × 5.25 inches × 1.2 inches, which is a comfortable size and weight to hold in the hand, and requires just a single AA-size battery.

Soon after, dedicated electronic stopwatches were produced but the ST-1 was the first combined with a calculator.

Casio MQ-1

Feature: Compact clock-calculator

LCD displays which could be powered from button cells became practical from around 1976 onwards. Casio already produced a number of miniature devices such as digital watches, calculator watches and calculators such as the Micro-mini, but in 1977 Casio combined a timepiece and a calculator and produced the MQ-1 (Micro-Quartz) calculator clock. The MQ-1 was a neat design at only 4.3 inches long and was powered by two button cells. Being a clock, it is always on and so a lid was fitted over the buttons to protect them from accidental presses whilst in your pocket. The lid also conveniently holds the operating instructions. It features a six-digit yellow-filtered LCD display, with time, date, clock and a stop-watch. It can also perform date calculations (between 1901 and 2099) taking leap years into account, i.e. calculate the number of days between two dates.

Later came a slightly larger MQ-2 version at 4.7 inches long × 1.4 inches deep. The MQ-2 improvements were an eight-digit display and the addition of an alarm.

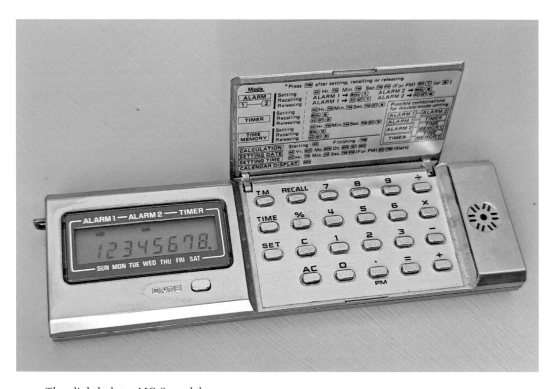

The slightly later MQ-2 model.

Kosmos 1

Feature: Biorhythm computer

This calculator from US company Kosmos International combined a calculator with a computer for working out what your day was going to be like in terms of your biorhythms. The analysis of a person's biorhythms was fashionable in the USA in the 1970s, and several companies produced devices to make the calculations. The theory, originated by Wilhelm Fliess in the nineteenth century, is that a person's life is governed by three cycles: a twenty-three-day physical cycle, a twenty-eight-day emotional cycle and a thirty-three-day intellectual cycle. By entering your personal details, it calculates where you are in each of the three cycles and predicts how you are going to feel on that day. Each of the three rhythms is the shape of a sine wave and a peak in the rhythm is supposed to be beneficial

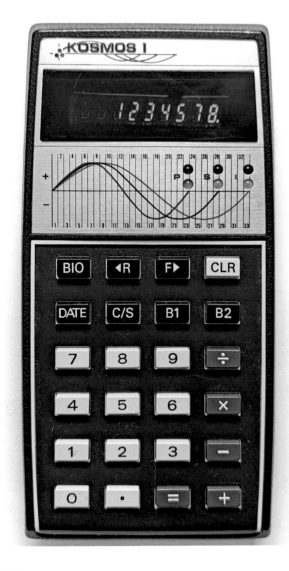

The Kosmos I biorhythm calculator.

for you in that area, i.e. physical, emotional or intellectual. Likewise, when the rhythm is in a trough for say, intellectual, it would not be advisable to appear on *Mastermind*! Of course, biorhythms are pseudoscience and despite studies made to try and link the theory to actual life events, no evidence was found. Kosmos also produced an astrology version called the Kosmos Astro for those who believe in horoscopes.

The calculator part is a four-function model with an eight-digit green fluorescent display and requires three AA-size batteries. They are not uncommon today on auction sites.

Casio SL-800 Film Card

Feature: Credit card-sized calculator
Although still rectangular, this calculator is included because its dimensions exactly mirror a credit card and was designed to fit into your card wallet. Produced around 1983, the size of 3.375 inches × 2.175 inches × 0.030 inches is a standard and is used for all credit and debit cards. There were many other calculators which matched the credit card height and width but only the SL-800 managed to achieve the proper thickness at the time. Being less than 1 mm thick and therefore having no room for a battery, power was from a solar cell. It had a memory, square root and per cent in addition to the basic functions, and also, being totally waterproof, it could be accidentally put through the washing machine without damage. It was a cool gadget at the time though, and the ultimate in miniaturisation!

The SL-760 model, a slightly thicker version of the SL-800.

Seiko C359-5000

Feature: Watch with a built-in calculator

In the early years of the 1970s, digital LED watches were the must-have gadget but suffered from poor battery life. A few years later, LCD versions were more practical being smaller, sleeker and with a battery life of many months rather than a few weeks. LCD watches with built-in calculators followed in the late 1970s and the Japanese watch manufacturer Seiko produced the C359-5000 in 1980. Seiko have a long history of watch manufacture dating back to 1881. Since then, they have achieved a number of firsts, including the first production quartz watch in 1969, the first TV watch in 1982 (yes, a TV screen on your wrist!), and the first automatic quartz watch in the late 1980s.

The C359-5000 can display the time in the twelve-hour format (AM/PM), and also the date and the day of the week. It has a twenty-four-hour alarm, a backlight and an hourly chime. The grill next to the buttons is the speaker for the alarm and the case and fine-link strap are stainless steel. The calculator buttons sit in two rows below the LCD and due to their tiny dimensions, require a stylus (included) or a ballpoint pen to work them. Unfortunately, when the stylus slipped off a button, it would scratch off the lettering above and so after not too long, all the buttons would look the same! By today's standards, the C359-5000 is a small watch: 1.35 inches (W) × 1.5 inches (L) × 0.37 inches (H), and weighs a lightweight 2.3 ounces. The back of the watch has a battery hatch which is opened by inserting a penny into the slot and twisting. This saves taking the back off and exposing the mechanism. There were a few different models based around the C359 case style, each with minor differences such as a black keyboard and a gold-coloured case. The C359 series are neat-looking watches and, even though not really practical as a calculator, still looks cool being worn today.

Seiko's watch-calculator combination.

This later Casio model – the CA-53 – has easier-to-use buttons.

Promotional Calculators

Over the years, many companies have used the calculator as way of advertising their products. They would ask a manufacturer to make a special batch with their name on them and then sell them or give them away with promotions. We have already seen a couple of examples of this with the Toshiba BC-0808 for Heineken and the Sinclair Sovereign for the Plessey company. Calculators eventually became so cheap that companies could afford to just give them away. The small cost would be more than covered by spreading their 'brand awareness' to the general public.

The following pages show some examples of these, starting with a couple from banks, promoting themselves by giving away credit-card-sized calculators with their names and phone numbers on the front.

Confectionery is an area where product promotion is big and the two elements of child education and their fondness for sweets come together. Nestlé produced a novel calculator in a chocolate-brown colour with buttons that look like smarties. Whether the intention was to introduce children to calculators or to suggest they buy some sweets after school is debatable.

On a similar theme, Mars gave away a calculator in the year 2000 inside a box of M&M's sweets. The calculator leaves no doubt as to what it is promoting with its M&M's characters, M&M's case-shape, M&M's-shaped buttons, and M&M's written on the front. Like the Nestlé calculator, it was also solar-powered and had the same functions – square root, per cent and a memory in addition to the basic four functions.

The final example is another from Nestlé, this one in the unmistakeable shape of a Polo mint. The Polo mint was first introduced in 1948 by Rowntree along with the tagline 'the-mint-with-the-hole'. This calculator dates to around 2007 and promotes the new Spearmint flavour.

Finally, no calculator book would be complete without showing a few words that can be written on calculators when viewed upside-down. The proper name for them is 'ambigrams' and an internet search will tell you that there are at least 250 ambigrams that can be written using the numbers: 8, 3, 6, 4, 1, 7, 0, 2 and 5 to represent the letters: B, E, G, H, I, L, O, S and Z.

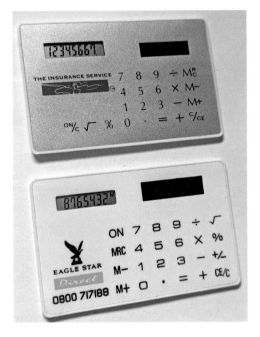

Give-away calculators from insurance companies.

Nestle's chocolate-inspired calculator.

Mars' colourful
M&M's calculator.

Front and rear views
of the calculator
with the hole.

How many of
these words do you
remember from
school?

Appendix
Maths Primer

Here is a brief explanation of some of the maths symbols and functions found on most scientific calculators.

$\sqrt{}$

The square root of a number x is the number which when multiplied by itself equals x. The square root of 4 is entered '4 $\sqrt{}$ ', and result = 2, because $2 \times 2 = 4$.

X^y (X to the power of y)

Calculates x multiplied by itself y number of times. To calculate 10^2, enter '10 X^y 2', result = 100. To find 5 cubed, enter '5 X^y 3', the answer = 125 (because $5 \times 5 \times 5 = 125$). If y is negative, the answer is the reciprocal of positive y, so 10^{-2} is 1/100 = 0.01.

Log (common logarithm)

Following on from above, if $z = X^y$, y is known as the logarithm of z for the case when x = 10. The number 10 is known as the base of the log. The log of 1,000 is therefore 3 (because $1,000 = 10^3$).

Ln (Natural Log)

Natural logs work the same way as common logs but instead of the base being 10, it is a constant, approximately 2.718, known by the symbol e. The natural logarithm of x is the power to which e would have to be raised to equal x. They may sound abstract but in fact they are very useful in many branches of maths, science and engineering.

e^x (exponential function)

The number e, also known as Euler's number, raised to the power of x is known as the exponential function. It is also the inverse of the natural log. The exponential function occurs so often in maths that mathematician Walter Rudin said that the exponential function is 'the most important function in mathematics'.

Sin, Cos, Tan (trigonometric functions sine, cosine and tangent)
These are usually introduced in schools as ways to calculate the angles in right-angled triangles when two of the sides of the triangle are known. For example, the sine of the angle is calculated as the length of the opposite side divided by the length of the hypotenuse. Then to find the actual angle, the inverse sine is applied, usually labelled as sin^{-1} on a calculator, which then gives the angle in whatever units are selected (see below). The inverse of sin, cos and tan may be labelled as ARC on a calculator. The trig functions are also used for solving many other problems in maths and engineering.

Deg, Rad, Grad (degrees, radians and gradians)
These are the three different units that angles can be displayed in. Radians are the most commonly used and a circle contains 2π radians. Many scientific calculators can display angles as degrees, minutes and seconds, known as sexagesimal (base 60) as there are 60 seconds in a minute and 60 minutes in a degree. Finally, gradians are the metric unit for angles, there being 100 gradians in a right angle.

Hyp (hyperbolic functions)
This button, used with the sin and cos buttons, gives the hyperbolic sine (sinh) and hyperbolic cosine (cosh) respectively. Where sin x and cos x form a circle when their points are plotted against each other, sinh x and cosh x form the right-hand half of a hyperbola curve. As well as being used in hyperbolic geometry, hyperbolic functions are used in many areas of maths and physics.

n! (factorial)
The factorial of a number n is defined as the product of all the positive numbers less than or equal to n. So the factorial of 4 is $4 \times 3 \times 2 \times 1 = 24$. They have many applications in mathematics including permutations (the counting of possible distinct sequences). The largest factorial a calculator can compute is 69. It is a good test of a calculator's performance to see how long the calculation takes – normally a second or two.

[(... 6 ...)]
Brackets (also known as parentheses) specify what order that a string of sums is performed in, e.g. '(((4 + 3) x (9 – 6)) – 2) / 3', tells the calculator to evaluate inside the brackets first, working from inside to outside. This example uses three levels of nesting, i.e. brackets within brackets, and many Casio's will let you have six levels of nesting.

σ (standard deviation)
Also written as σn or abbreviated to SD, standard deviation is a measure of the amount of variation of a set of values. A low standard deviation indicates that the values tend to be close to the mean of the set, while a high standard deviation indicates that the values are spread out over a wider range. Calculators with statistical functions require you to enter the set of values first then calculate the SD, or other stats like the mean or σn -1, known as the sample standard deviation.

Acknowledgements

Thanks to vintagecalculators.com for permission to use their photos of the Busicom calculator.

About the Author

Andrew Morten had an interest in electronics from an early age and went on to gain his degree in Electronics and Electrical Engineering from Coventry in 1987. He then spent his career working on various projects in both the industrial and defence sectors, at companies such as Plessey, Racal and General Electric. He is now retired and lives in Leicestershire.